ESTIMATING
Home Construction Costs
—— SECOND EDITION ——

ESTIMATING
Home Construction Costs
── SECOND EDITION ──

Jerry Householder
with Emile Marchive III

Estimating Home Construction Costs, Second Edition

BuilderBooks, a Service of the National Association of Home Builders

Courtenay S. Brown	Director, Book Publishing
Doris M. Tennyson	Senior Editor
Natalie C. Holmes	Book Editor
Torrie Singletary	Production Editor
Studio Grafik	Cover Design
Circle Graphics	Composition
King Printing Co., Inc.	Printing

Gerald M. Howard	NAHB Executive Vice President and CEO
Mark Pursell	NAHB Senior Staff Vice President, Marketing & Sales Group
Lakisha Campbell	NAHB Staff Vice President, Publications & Affinity Programs

Disclaimer

This publication provides accurate information on the subject matter covered. The publisher is selling it with the understanding that the publisher is not providing legal, accounting, or other professional service. If you need legal advice or other expert assistance, obtain the services of a qualified professional experienced in the subject matter involved. Reference herein to any specific commercial products, process, or service by trade name, trademark, manufacturer, or otherwise does not necessarily constitute or imply its endorsement, recommendation, or favored status by the National Association of Home Builders. The views and opinions of the author expressed in this publication do not necessarily state or reflect those of the National Association of Home Builders, and they shall not be used to advertise or endorse a product.

Printed in the United States of America

21 20 19 18   3 4 5

ISBN-13 978-0-86718-615-4   ISBN-10 0-86718-615-1

Cataloging-in-Publication information available on request.

For further information, please contact:

National Association of Home Builders
1201 15th Street, NW
Washington, DC 20005-2800
800-223-2665
Visit us online at www.BuilderBooks.com.

Contents

Figures

Acknowledgments

The authors and NAHB BuilderBooks, wish to acknowledge the contributions of the following reviewers who assisted with reviews of the proposal or the developing manuscript of the first edition: Ken Cameron, Mercury Homes, Aurora, Illinois; Scott Condreay, Purdue University, West Lafayette, Indiana; Robert Deppe, Robert Deppe, Inc., Caledonia, Michigan; Leslie Feigenbaum, Texas A and M University, College Station, Texas; John Fredley, University of Washington, Seattle, Washington; Robert Hankin, RDR Construction Corporation, Highland Mills, New York; Allan Freedman, NAHB Business Management, Washington, DC; Nancy Holland, Texas A and M University, College Station, Texas; Pat Jarrett, Homes by C.P. Morgan, Carmel, Indiana; Richard Langedizk, Construction Estimating Institute, Sarasota, Florida; Tom Love, Doster Construction Company, Irondale, Alabama; Mike Maxwell, Purdue University, Indianapolis, Indiana; Russel Neumann, Vista South Custom Homes, Madison, Alabama; Clark Pace, University of Washington, Seattle, Washington; John Piazza, Piazza Construction Inc., Mt. Vernon, Washington; David Roberts, David R. Roberts Builder, Inc., Kalamazoo, Michigan; Leon Rogers, Brigham Young University, Provo, Utah; Paul Scholes, David Weekly Homes, Houston, Texas; Matt Syal, Michigan State University, East Lansing, Michigan; and Bobby Whitten, Home Builders Advisory, Madison, Alabama.

About
the Authors

Jerry Householder, builder, professor, and author, has many years of experience as a general contractor. He has built over $100 million worth of residential, commercial, and industrial projects throughout the southeastern United States. A professor for over 20 years, he holds a doctorate in civil engineering from the Georgia Institute of Technology. Dr. Householder is a retired Distinguished Professor at Louisiana State University and continues to build homes.

Emile Marchive III is an instructor in the Construction Management Department at Louisiana State University in Baton Rouge. He holds a master's degree in Building Science from Auburn University and is scheduled to earn a doctorate in Engineering Science from LSU in 2006. He has been active in his family's construction business for nine years.

Estimating Methods

Estimating is the process of determining the quantities of work and materials required and the probable cost of a home. A builder analyzes the many different elements that go into building a home and how much to buy when forecasting the costs. Like most other ventures, the value or accuracy of the estimate is directly proportional to the effort put into the task. In other words, the more work a builder puts into an estimate, the closer the forecasted costs are likely to be to the actual costs. This book is intended for builders and others with some basic knowledge of home construction. While some estimating experience is helpful, it is not essential to successfully apply the principles explored in this book.

Purposes of Estimating

An estimate serves many purposes and helps the builder in all phases of his or her work. Different kinds of estimates are available, the kind you choose to do will depend on the purpose of the estimate.

Determine the Price for a Job

A builder has many reasons for forecasting the price of a home. One of those reasons is to determine what his or her cost should be. The feasibility of a project usually depends upon its cost. Sometimes a builder needs an accurate forecast of the cost and sometimes a less accurate estimate of the cost will suffice. For example, a speculative builder may consider several alternative plans with similar market values and would like a rough idea of each plan's cost. In this situation the builder might begin by narrowing the potential plans to two or three by using less accurate but quicker estimating methods. In such a situation, if a builder is considering five different plans, all of which are suitable for a particular location, he or she might use one of the less detailed methods discussed later to choose two sets of plans. The builder then would use a more detailed and accurate method to estimate the probable cost of these two selections in order to make the final choice.

If a builder is negotiating with an owner to build a custom home, the builder might give the owner approximate costs for various plans that the owner is considering. By giving a quick, ballpark price, the builder can help clients make better decisions about which plan or features best fit their needs and budget. However, before giving a final, fixed price, the builder should perform a more detailed and accurate cost estimate. Likewise, if the home is to be built on a cost-plus basis where the owners pay the builder's cost plus an amount for profit and overhead, a detailed estimate is necessary to establish a reasonable target budget.

When a builder submits a bid to an owner offering to build a home for a fixed price, the builder needs to know, within a reasonable degree of probability, what the project should cost. Although some builders may make a firm offer without first doing a detailed estimate, experience has shown that this practice is dangerous.

If you, as the builder, are financing the construction, you must have a good idea of its cost. Bankers feel more secure when a builder demonstrates sound and careful planning. By showing a banker a detailed estimate, a sound critical path method (CPM) schedule, and a reasonable market study, you will enhance your chances of getting financing.

Establish a List of Materials

On a typical construction project, the builder furnishes some of the materials, and the subcontractors furnish others. The custom and practice of who furnishes what may vary from region to region. Plumbers and electricians typically furnish most of their materials; heating and air-conditioning contractors usually furnish all of their materials. Other subcontractors such as painters, landscapers, drywallers, floor covering subcontractors, and insulators typically furnish some or all of the materials they install. Generally, in today's construction environment, the builder provides all materials used by framers, roofers, and masons. When performing a detailed estimate, the builder must establish who will furnish various materials.

When doing one of the more approximate estimates, the builder does not "take off" or estimate the quantity of materials. However, when doing a detailed estimate, the builder typically determines the quantities of the materials that he or she will furnish. A detailed estimate, such as the one shown in later chapters of this book, can be used during construction to order most of the materials.

Use the Estimate to Help in Design and Marketing

Designing a home involves making comparisons and choices. Except in cases where a client presents a set of plans and specifications that are totally complete, you will typically make many functional design decisions. If one of your major goals is to maximize profits for a given work effort, you need to make informed design decisions.

From time to time and place to place, different design features dramatically affect the marketability of your homes. Determining which features will be most desirable to your potential buyers is a matter of art. Your own experience, seeing what sells for other builders, and the experience of realtors can help you in this regard. A word of caution—just because you personally place great value on a particular design feature does not necessarily mean that potential buyers will.

The first aspect of making design choices, that of determining how highly buyers will value certain design features, is often a matter of subjective opinion. The second aspect, that of comparing the costs of these alternatives, is not. For example, assume that you are trying to decide between putting a steeply pitched roof with dormers on a house or a less complicated and less expensive roof. You can do a fairly detailed estimate of the cost of each roof and determine the difference. Once you know how much more one costs than the other, your final decision will be easier.

Scheduling

Much has been written about the benefits of doing CPM schedules for home construction, and most builders who use modern scheduling techniques will attest to those benefits. A detailed estimate can help you develop a workable schedule.

When you do a CPM schedule, the first step is to study the plans and ascertain the different activities necessary to do the job. Reviewing a detailed estimate will help in this effort because the estimator will often choose the methods and techniques that will be used. The second step is to determine in what order the work will be done— that is, what activities will precede, be concurrent with, or follow others. The estimator must of necessity determine how the work will be done, what equipment will be necessary, and what special considerations must be taken into account. The scheduler and estimator must agree on these points. Finally, the scheduler must determine the duration of the different activities. The estimate helps because it lists the quantities of material and worker-hours for many activities.

Job Cost Control

Job cost control involves tracking costs as they occur and comparing them to the budget. The budget for a job is simply the expected cost of each category of work such as sitework, masonry, and framing. If the categories are broken down so that they are completed chronologically as the job progresses, you can compare your actual costs of most work categories to their budgeted costs before the whole job is finished. In this way you can determine whether you are over or under budget. When the project is over, you will be able to easily compare your estimated costs with the actual costs. If the estimate is an organized one, with the work listed in the appropriate category, it can easily be used to determine the cost to complete each category and the job as a whole. If you compare costs and determine the reasons for any variances, you can determine the weaknesses of your estimating system and hopefully improve it. Over time, if you compile and update an historical database of costs by category, you will make your estimates more reliable.

Establishing a budget for an entire house and for each individual work category is the businesslike way of running any building operation. In fixed price work, the builder needs to track the cost of each separate category and compare it to the budget for that category to determine whether the project is within budget. In cost-plus work, the builder should try to keep the owner's best interest at heart. This effort requires tracking the costs and comparing them with a definitive budget by category using a detailed estimate.

Communication

The detailed estimate not only tells the person responsible for purchasing materials what to order, but it also tells the person in charge of field construction what methods and equipment the estimator had in mind when the job was estimated. You should note the names of the trade contractors whose bids you used to compile the estimate on the estimate itself. This practice may be particularly important to builders who use a variety of trade contractors on their jobs. While builders gain certain advantages by using the same trade contractors over and over, they derive other advantages from using different trade contractors.

Usually, the person in the field who is in charge of construction has the option of changing the methods of performing the various tasks that are indicated in the estimate. An experienced superintendent will look for ways to save the company money. However, in getting lower trade contract bids, certain ethical considerations need to be considered. One school of thought is that the builder should take these bids and choose the best one without further negotiation. A second school of thought is that procuring the services of trade contractors is no different from obtaining any other service or commodity on the open market. Builders who subscribe to this latter philosophy often try to negotiate their trade contractors down on their prices or reveal the lowest bid to other trade contractors in order to get a better price. The problem for a builder who gets the reputation of being a bid shopper is that trade contractors will eventually start out higher than their best price in preparation for the forthcoming negotiation. Many people question whether bid chiseling or bid shopping actually leads to better overall prices.

Types of Estimates

You can estimate the probable cost of a house many ways. At one extreme, a builder may simply look at the plans and, without doing any calculations, come up with the forecasted cost. If a builder is sufficiently experienced with similar homes, he or she might be able to guess the cost within 10 to 15 percent. At the other extreme, a builder can estimate the probable cost of a home by determining the required quantities of each and every item needed, down to the last nail, and then multiply them by firm material quotes. If you add this amount to labor and equipment costs determined from established, historical production rates and the firm quotes from trade contractors, you can determine the probable cost. Between these two extremes are any number of methods. Three of these methods are discussed in this book: the square-foot method, the component or assemblies method, and the detailed method. Each type of estimating serves a different purpose.

Square-Foot Method

Of all the methods used to determine the cost of a home, the square-foot method is the most well known. Real estate agents, potential home buyers, appraisers, courts of law, and builders all utilize this pricing method. What is not widely understood, however, is the degree to which this method may be inaccurate. If the size, quality of materials, number of baths, quantity of cabinets, and amenities of two different homes built at the same time in the same area are similar, then their square-foot costs should likewise be similar. But if any of these variables differ, then the square-foot costs will probably differ.

Given the fact that real estate agents, appraisers, and potential buyers use the square-foot method of estimating to establish the value of a home, a builder can maximize profits by using a little common sense. For example, assume that you are building in an area where all the homes have similar design features and are of somewhat similar size. Since the square footage of living area and the sales price of a home are easily discoverable, a square-footage cost based on total sales price (including the land) divided by the living area (not counting porches, garage, and outside storage) is commonly used to determine the value of your home. A smart builder can construct a home with slightly larger rooms and a slightly smaller porch and make more profit than if the rooms and porch are similar in size to the competition. Adding features that cost more without adding area can be risky. If you add a $3,000 fireplace to a 3,000 square-foot home, you have just added

$1 per square foot to the cost. If a buyer is more cost conscious than quality conscious, you may not recover all of this cost. Furthermore, if you use expensive features where more moderately priced ones will serve the same purpose, many buyers are not likely to appreciate the difference. For example, if you choose windows that cost $500 each, you may not be able to recover the difference over using windows that cost $200 each.

So you can see that if real estate agents, appraisers, and buyers use the square-foot method to determine the value of your home, you can increase your profits by choosing designs and materials that increase the value when measured in this way but do not proportionately increase the cost.

A second use of square-foot estimating is to estimate what your costs should be. The accuracy of the square-foot method increases when the house you are building and all the other building conditions are similar to a house for which you have cost records. The more similar they are, the more confident you can be that the resulting estimate will be reasonably close to the actual cost. For houses that are less similar, the square-foot method may be useful to come up with a quick, ballpark guess in situations where various plans are being considered.

You should always figure the square-foot cost of houses on which you perform a detailed estimate as a check against potential blunders. To do this calculation, simply divide the cost as estimated by the detailed method by the square footage and compare the square-foot cost with other homes you have built and estimated. If the cost seems too high or low, check your detailed estimate for mistakes.

When using the square-foot method to estimate the cost, you need to do it in a way that gives you the most accurate results. In determining the cost per square foot, you divide the cost by the square footage. However, you need to determine which particular cost and which particular square footage numbers to use. Consider the following situation:

Cost of house	$300,000	Living area	3,000 sq. ft.
Cost of lot	$100,000	Nonliving area	1,000 sq. ft.
Sales Price	$500,000	Total area	4,000 sq. ft.

The square-foot cost may be the cost to the buyer, $500,000, divided by the square foot of living area, 3,000.

$$\frac{\$500,000}{3,000 \text{ sq. ft.}} = \$166.66/ \text{ sq. ft.}$$

The cost of construction per square foot of total area is the net cost of the house without land, $200,000, divided by the total area, 4,000 square feet.

$$\frac{\$300,000}{4,000 \text{ sq. ft.}} = \$75.00 / \text{ sq. ft.}$$

Compare these numbers to a house whose cost you do not know, which has these areas:

Living area	3,500 sq. ft.
Nonliving area	500 sq. ft.
Total area	4,000 sq. ft.

In this house, everything is the same except that the finished living area is 500 square feet larger and the nonliving area (porches, garage, and storage space) is 500 square feet smaller. If you use $75 per square foot of total area to estimate the cost, you will come up with—

$$\$75 \text{ per sq. ft.} \times 4,000 \text{ sq. ft.} = \$300,000$$

However, this house obviously will cost more to build than the previous house if the quality of materials are the same because you are finishing off 500 more square feet.

What you need to do when using the square-foot method to determine your cost is to pull out the variable items that you can price separately, such as the land cost. Furthermore, you need some way to take into account the effect of the nonliving area. One way to accomplish this is to figure an effective or modified square footage. To do this calculation, estimate how much it takes to build nonliving areas such as porches, garages, and basements as a percentage of the finished living areas. You may find that for the homes that you are building, the nonliving areas cost 40 percent, 50 percent, or even 60 percent of the cost of the living area.

Example—The nonliving area costs you 50 percent of what it costs to build the living area. In this case, to determine the square-foot cost base on the effective area, you figure the effective area, which is the living area plus 50 percent of the nonliving area. For example, in the house with 3,000 square feet of living area and 1,000 nonliving area, the effective area would be 3,000 + (50% × 1,000) or 3,500 square feet. If the house without the land cost $300,000 to build, the cost per square foot of effective area is $85.71. You can then use this number to estimate the cost of the 3,500 square-foot house with 500 square feet of nonliving area by using the equivalent area of 3,500 + (50% × 500) or 3,750 square feet. The estimated cost of this house would be 3,750 × $85.71 = $329,984.

Component Method

The component method of estimating is sometimes called the conceptual method or the assemblies method. When using this method, the builder determines the probable cost of several different components or assemblies of the house and adds them together. The first step is to determine the lineal feet of exterior walls, the lineal feet of interior walls, and the square footage under roof. Multiply these numbers by the unit cost of each and add the costs of the other components such as plumbing, electrical, and cabinetry to determine the final cost.

As in any estimating procedure in which the costs of the different elements of the project are added together, the greatest chance for error comes from leaving something out. A checklist that includes all the items that add to the cost helps ensure the estimate's accuracy.

You will need to establish your own checklist based on your operation. One suggested way is as follows:

1. Land costs
2. Exterior walls
3. Interior walls
4. Square feet of covered area
5. Windows and doors
6. Plumbing; heating, ventilation, and air-conditioning (HVAC); and electrical
7. Sitework
8. Fireplaces and stairs
9. Kitchen cabinets and appliances
10. Exterior work
11. Job overhead

Land Costs

This component includes the raw cost of the land plus any cost associated with water supply, sewage disposal, and offsite utilities. By including the water, sewer, and utility costs, the builder can more easily compare different land costs since some building lots come with water and sewer available at no extra cost while other sites do not. However, if the builder is developing land from scratch, this approach is not sufficient. In this instance, you will have to consider all development costs separately and determine the per-lot cost prior to estimating the cost of each home.

Exterior Walls

You will need to calculate the cost of one lineal foot of exterior wall for the plan that you are estimating. Refer to the wall section shown in the plans, figure the cost of each lineal foot of wall, and multiply this number by the lineal footage. For most houses, you would figure the total footage and subtract openings greater than 4 lineal feet. By covering up smaller openings, you can account for waste. Be sure to figure all the costs including footings and overhangs:

- footing costs—concrete, rebar, formboards, labor
- 1 lineal foot sill plate
- 1 stud
- 2 lineal feet top plate
- wall sheathing (8 square feet for an 8-foot wall)
- drywall—hung, finished, and painted, (8 square feet for an 8-foot wall)
- 1 lineal foot base
- insulation (8 square feet for an 8-foot wall)
- crown molding, 1 lineal foot
- 1 lineal foot, roof overhang, including—
  - rafter tails
  - lookout
  - plywood roof decking
  - roofing and labor for overhang
  - facia and subfacia
  - soffit
  - frieze
  - gutter
  - 1 lineal foot exterior skin (brick, siding, and the like)

If the walls of the home you are estimating have more than one covering, for example, some are brick and some are siding, you will need to figure the lineal foot cost of each.

Interior Wall

This component includes—

- 1 lineal foot sill plate
- 1 stud
- 2 lineal feet top plate
- drywall—hung, finished and painted (16 square feet for an 8-foot wall)
- 2 lineal feet base
- 2 lineal feet crown molding

Square Foot of Area

This component includes all the costs of a 1-square-foot column from the middle of a room. Items may include:

- 1 square foot floor
  - 1 square foot concrete slab
    - 1 square foot gravel
    - 1 square foot polyethylene
    - 1 square foot welded wire mesh
    - 1 square foot concrete and placement labor
  - Framed floor
    - 1 lineal foot floor joist
    - 1 square foot insulation
    - 1 square foot plywood decking
  - 1 square foot floor covering
  - 1 square foot finished drywall for ceiling
  - 1 lineal foot ceiling joist
  - 1 square foot insulation
  - 1-plus* lineal foot rafter
  - 1-plus* square foot plywood
  - 1-plus* square foot roof felt
  - 1-plus* square foot roofing and labor
  - Subcontract costs that are charged by the square foot, for example—
    - slab-forming labor
    - framing labor
    - trim carpentry labor
    - painting labor

Windows, Doors, and Attachments

Windows, doors, hardware, and skylights are priced separately. This category also includes any costs associated with the exterior skin not accounted for elsewhere, such as gable ends and dormers.

Plumbing, HVAC, and Electrical

In addition to the base trade contract prices, include items you will furnish such as fixtures, mirrors, tub and shower enclosures, and bath accessories.

Site Work

Include any costs to prepare the site such as rough grading, fill, leveling, and compaction.

Fireplace and Stairs

Add all costs associated with these items.

Cabinets and Appliances

Also include any special cabinetry not in the bath or kitchen.

Exterior Work

This category includes—

- final grading
- landscaping

---

*Each of these depends on the roof pitch.

- walks and drives
- irrigation and sprinkler systems
- decks
- fencing
- general cleaning for the entire project

Job Overhead

This category includes all those items that are not chargeable to a particular work item but are still chargeable to the job including—

- plans
- loan closing costs
- permits and fees
- insurance
- construction loan interest
- superintendent
- toilets
- job-site office
- sales costs

A more detailed checklist for closing costs appears in Figure 2.2.

Detailed Estimate Method

The detailed or complete estimate is more accurate than the square-foot estimate or the component method; however, the detailed estimate takes more time. When using the detailed estimating method, the builder must determine—

- all the different items such as materials, labor, trade contracts, and equipment that are required
- how many of each item are needed
- the price of each item

In the detailed method, the builder divides the work into categories. For commercial building construction, the work is usually divided into the 16 divisions of the Construction Specifications Institute (CSI) format. Each builder must determine how best to subdivide his or her own work. The examples used in this book divide the work into 18 categories. The categories are further broken down into individual items to be quantified and estimated. When developing your own checklist, establish logical categories with activities that will be finished at about the same time. This arrangement helps when you are comparing your costs to the budgeted amount as the job progresses.

One of the most important tools that an estimator can have is a complete checklist. The checklist helps organize the estimate and prevent the omission of any necessary items. A sample checklist appears in Figure 2.2 in Chapter 2.

A detailed estimate differs from other, more approximate estimating methods primarily because the detailed estimate involves determining all the quantities or work and materials required. A quantity or material takeoff, is an itemized listing of the materials needed to build the house. After the quantity takeoff is finished, the next step is pricing the work. Finally the prices of each category are added up or recapped to determine the direct cost of the job. Adding profit and general overhead gives you the total cost.

Except on the occasion when the house is already finished and the cost is known or the builder has built a similar house under similar circumstances, you should not offer to build a house for a fixed price without performing a detailed estimate. The survival of builders who do fixed price work depends upon the accuracy of their estimates. Therefore, you should use the most accurate method possible to determine what the cost is likely to be.

2

Quantity Takeoff

The detailed method of estimating must offer extensive review and control capabilities. The first major step in performing a detailed estimate is to make a quantity takeoff. The quantity takeoff or quantity survey is the list of all the materials that the builder must provide and other items of work that he or she must count. If the takeoff is done in an appropriate manner, it may be used during construction to order the materials. Pricing the individual work items is a separate step performed after the quantity takeoff is complete.

Preliminary Considerations

You should record on worksheets or quantity sheets all the calculations and assumptions that you use in arriving at your quantities. Notes should indicate the location of the item, the trade with which it is associated, and any conversion factors and formulas you used. There are many reasons for recording your calculations so that someone unfamiliar with the project can follow your logic. If your quantities turn out to be wrong during construction, you can more easily review your work to find the mistake so that you will not make it again. In organizations of more than one person, sometimes others need to use, review, or modify your takeoff. If you are required to change the scope of work during its progress, you can more easily adjust your material quantities if you have a record of exactly how you arrived at your numbers. Quantity takeoffs are usually performed on quantity worksheets such as the one appears in Figure 2.1. Using a worksheet such as this one helps to keep the takeoff organized.

Familiarization with the Plans

The first thing you do when performing a quantity takeoff is to thoroughly study the plans and specifications and visualize the job being built. Make notes of any particular procedure or method of construction that you plan to do that would not be

FIGURE 2.1 Quantity Takeoff Form

CATEGORY:					QUANTITY SHEET				DATE	
ESTIMATOR:									PAGE:	
JOB:										
DESCRIPTION	NO.	DIMENSIONS			SUBTOTAL		TOTAL		REMARKS	
					QUANTITY	UNIT	QUANTITY	UNIT		

obvious to someone else in your organization. Second, be on the lookout for materials or work items that are not listed on your checklist. If the house is a custom home being built for an owner, carefully read the general and supplemental conditions to the contract to see if it contains any unusual requirements that might affect how you price the work.

Site Visit

You should not estimate the cost of a house or do a quantity takeoff without performing a site visit in order to understand site conditions that may require additional work. The kind of information that you should get during the site visit includes—

- topography of the site
- existence of trees and other vegetation
- low areas and soft spots
- probability of subsurface rock or water
- access to the site and access around the site
- availability of utilities
- storage areas
- necessity of protecting adjacent property

You should prepare a checklist in advance to help you in your investigation of the site.

Account for Waste

When you walk on a construction site and look at the trash to be hauled away, you will see the usual assortment of cartons, wrappers, and other products not directly associated with the work. You will also find short pieces of lumber, cut up drywall, broken brick, pieces of roofing shingles, and other wasted material. Such waste is inevitable and must be accounted for when you are estimating quantities. You can account for waste in an estimate in several ways. One is to figure the quantities and then add a waste factor in the form of a percentage. For example, when you take off the brick, you may find that if you order the exact in-place quantity you will not have enough to do the job. In this case you will need to add some percentage, say 5 percent, to cover the waste.

A second way to allow for waste is to include openings when taking off quantities. For example, when you take off drywall, if you figure enough material to cover all openings with widths of 4 feet or less, you will usually have enough for waste. Variations of these methods for individual materials are discussed later in this chapter.

Perform the Takeoff

Establish and Use a Checklist

Many estimators believe that the most important element of a proper takeoff is identifying and listing every element so that nothing will be left out. As mentioned earlier, you should develop a checklist to which you can refer when doing the quantity takeoff and when pricing the work.

The best checklist is one that you make based on your work. The items can vary from one checklist to another depending on the price range of the house being built, local practices, material availabilities, and builders' preferences. The checklist in Figure 2.2 is general and cannot possibly be perfect for every single builder. However, it can serve as a starting point if you don't have one already. If you do have one, you can compare the two to help make yours more nearly complete.

As you go through the estimate, you will be applying your own methods to estimate the various quantities. Following are some commonly used methods to estimate the quantities that are found in home construction.

Land Costs

This book does not cover how to estimate land development costs. If you are building on lots that you have developed from scratch, the land cost must include your total cost on a per-lot basis. The reason that well, septic system, and offsite utility costs are listed is that in most situations these amenities are already in and are a part of the land cost. If they are not, as is the case with many rural lots, these costs need to be added to the raw land cost so that you can make a direct comparison.

Site Work

When you make your site visit, you should look, among other items, for the site work you will have to do. If trees must be removed, you need to note how many truck loads you will have. If any trees must be taken down by an experienced tree-cutting company because of their proximity to power lines or other houses, you need to note how many of those there are.

Excavation and fill may be divided into three classifications for estimating:

- general reshaping of the site by cut and fill
- excavation of holes for basements and/or deep footings
- backfill

Your contract may require general reshaping of the site, or you may decide to move the dirt around, bring some in, or take some away. Some builders will look at the site and just "eyeball" the amount of work to be done. If you have a lot of experience or are lucky, this technique might be okay. If you want a more precise estimate, you will need a topography map of the existing and final grades. Most of the time when you do general grading, both cut and fill will be necessary. You need to separate the two quantities in the estimate. To do this separation, start with your topographical map that shows existing and final grades and draw a line across it to where the cut changes to fill. On one side of the line is all cut, and on the other side is all fill. In doing a general grade reshaping, you can often reuse the existing topsoil. If you can remove and stockpile the topsoil, be sure to include this grading operation in your calculations. To estimate the quantities of cut and fill, draw a grid on your map that has the combined before and after topography. Most estimators who perform this kind of estimate regularly use a 25-foot or 50-foot grid. After you have drawn the grid, calculate the existing and final elevations of the ground at each grid point. You then calculate the amount of cut or fill within each square as established by your grid.

You can do this calculation two ways depending on whether the soil within a grid is entirely cut, entirely fill, or both cut and fill. If the area within a grid is entirely cut or entirely fill, simply added the depth of cut (or fill) at each corner and divide by four to get the average and then multiply this average by the area of the grid.

Example—You have drawn a 25-foot by 25-foot grid and have determined that the amount of cut at each of the four corners is 2.0 feet, 3.1 feet, 4.2 feet, and 5.1 feet, you add these four numbers and divide by four. The average cut in this case is 3.6 feet. Multiply 3.6 feet by the area of the grid, 25 feet by 25 feet or 625 square feet to get

FIGURE 2.2 Estimating Checklist

1. Land Costs
- ☐ Raw land costs
- ☐ Well or water system
- ☐ Septic systems
- ☐ Offsite utility costs

2. Job Overhead
- ☐ Architectural and engineering
  - ___ Plans
  - ___ Surveys
- ☐ Land closing costs
  - ___ Appraisal
  - ___ Discount points/origination fee
  - ___ Title examination
  - ___ Document preparation
  - ___ Recording fee
  - ___ Attorney's fee
  - ___ Photographs
  - ___ Taxes
- ☐ Construction loan closing
  - ___ Appraisal
  - ___ Discount points/origination fee
  - ___ Title examination
  - ___ Document preparation
  - ___ Recording fee
  - ___ Attorney's fee
- ☐ Permits and fees
  - ___ Building permits
  - ___ Home builders association assessment
  - ___ Warranty fees
  - ___ Inspection fees
- ☐ Impact fees (school, park, etc.)
  - ___ Tap fees and utilities
  - ___ Water tap fees
  - ___ Sewer tap fees
  - ___ Temporary power hookup
  - ___ Temporary utilities
  - ___ Water usage
- ☐ Special environmental fees
- ☐ Insurance
  - ___ Builder's risk insurance
  - ___ Liability insurance
  - ___ Completed operations insurance
- ☐ Interest on construction loan
- ☐ Sales cost
  - ___ Appraisal
  - ___ Discount points/origination fee
  - ___ Title examination
  - ___ Document preparation
  - ___ Recording fee
  - ___ Attorney's fee
  - ___ Photographs
  - ___ Taxes
  - ___ Sales commissions
  - ___ Special warranty costs

- ☐ Site overhead expenses
  - ___ Job superintendent
  - ___ Guard costs
  - ___ Temporary toilets
  - ___ Fences
  - ___ Jobsite office
  - ___ Storage facilities
  - ___ Small tools

3. Site Work
- ☐ Lot clearing
  - ___ Site Clearing
  - ___ Fill material/hauling
  - ___ Rough grading
  - ___ Demolition and disposal
  - ___ Culverts
  - ___ Erosion control
  - ___ Compaction

4. Footings and Slabs
- ☐ Batter boards
- ☐ Formwork
  - ___ Formboards
  - ___ Brick ledge
  - ___ Stakes
  - ___ Braces
  - ___ Form labor
  - ___ Special forms
- ☐ Dig footings
- ☐ Reinforcing steel
  - ___ Dowels
  - ___ Anchor bolts
  - ___ Rebars
  - ___ Tie wire
  - ___ Wire mesh
  - ___ Placement labor
- ☐ Porous fill
  - ___ Porous fill material
  - ___ Installation labor
- ☐ Termite and soil treatment
  - ___ Soil poisoning
  - ___ Termite shields
  - ___ Radon control
- ☐ Vapor control
  - ___ Vapor barrier material
  - ___ Placement labor
- ☐ Concrete
  - ___ Materials
  - ___ Placement and finish labor
  - ___ Curing and protection
  - ___ Pumps, vibrators, and other equipment

5. Masonry
- ☐ Masonry units
  - ___ Brick
  - ___ Block
  - ___ Stone
  - ___ Masonry lintels
  - ___ Special masonry

- ☐ Fireplace
  - ___ Prefab unit/chimney
  - ___ Firebrick
  - ___ Damper
  - ___ Steel angles
  - ___ Flue liners
  - ___ Common brick
  - ___ Chimney cap
  - ___ Miscellaneous equipment
- ☐ Mortar
  - ___ Mortar mix/cement
  - ___ Sand
  - ___ Color
  - ___ Admixtures
  - ___ Lime
- ☐ Metal
  - ___ Wall ties
  - ___ Metal lintels
  - ___ Wall reinforcing
- ☐ Waterproofing
  - ___ Waterproofing materials
  - ___ Drainage materials
  - ___ Piping
  - ___ Waterproofing labor
- ☐ Stucco
- ☐ Scaffolding
- ☐ Masonry labor
- ☐ Masonry equipment

6. Framing
- ☐ Floors
  - ___ Sill plates
  - ___ Sill sealer
  - ___ Floor trusses
  - ___ Wood beams
  - ___ Steel beams
  - ___ Joist hangers/ledgers
  - ___ Joists and bands
  - ___ Bridging
  - ___ Sleepers for concrete
  - ___ Subfloor
- ☐ Nonmasonry piers
- ☐ Walls
  - ___ Treated plates
  - ___ Untreated plates
  - ___ Studs
  - ___ Headers
  - ___ Gable framing
  - ___ Blocking
  - ___ Corner bracing
  - ___ Sheathing
  - ___ House wrap
  - ___ Posts/beams
  - ___ Garage beams
- ☐ Roof
  - ___ Trusses
  - ___ Ceiling joists
  - ___ Rafters
  - ___ Ridge members

FIGURE 2.2 *Continued*

6. Framing (*continued*)

___ Hip and valley members
___ Bracing and strongbacks
___ Subfacia/lookouts/band
___ Roof sheathing/plywood clips
___ Felt/ice dam
☐ Exterior millwork
   ___ Siding
   ___ Facia
   ___ Soffit and garage ceiling
   ___ Porch ceiling
   ___ Frieze
   ___ Louvers and vents
   ___ Exterior locksets
   ___ Windows/skylights
   ___ Window wrap
   ___ Trim boards
   ___ Exterior doors and
       threshholds/storm doors
   ___ Soffit vents
☐ Stair framing
   ___ Stringers
   ___ Risers
   ___ Treads
☐ Decks
   ___ Posts
   ___ Beams
   ___ Joists
   ___ Blocking
   ___ Decking
   ___ Rail materials
   ___ Stairs
   ___ Joist hangers
   ___ Deck foundations
☐ Miscellaneous carpentry
   ___ Fences
   ___ Lattice
   ___ Shutters
☐ Fasteners
   ___ Nails
   ___ Bolts
   ___ Adhesive
   ___ Screws
   ___ Column bases
☐ Framing labor

7. Roofing

☐ Roof edge
☐ Valley material
☐ Roofing materials
   ___ Asphalt shingles
   ___ Wood shingles
   ___ Tile
   ___ Metal roofing
   ___ Slate
☐ Flashing
☐ Nails and fasteners
☐ Built-up/single-ply roofing

☐ Ridge vents/roof vents
☐ Roofing labor

8. Plumbing

☐ Basic plumbing subcontract package
☐ Vanity tops
☐ Tubs
☐ Sinks
☐ Hot water heater
☐ Shower
☐ Tub enclosures
☐ Shower doors
☐ Sewer and water tie-in
☐ Fireplace gas

9. Electrical

☐ Basic electrical subcontract package
☐ Fixtures
☐ Telephone wiring
☐ Cable television
☐ Sound system
☐ Burglar alarm
☐ System control (smart house)

10. Heating, ventilation, and air-conditioning (HVAC)

☐ Basic HVAC subcontract package
☐ Attic fans
☐ Ventilation equipment
☐ Air-conditioning pads

11. Insulation

☐ Attic/ceiling
☐ Walls
☐ Slab/basement walls
☐ Rigid board
☐ Special insulation
☐ Insulation labor
☐ Baffles

12. Drywall

☐ Regular drywall
☐ Moisture resistant drywall
☐ Fire code drywall
☐ Nails, screws, tape, joint
   compound, glue, corner bead
☐ Installation labor
☐ Finishing labor

13. Interior Trim

☐ Moldings
   ___ Base/shoe molding
   ___ Crown
   ___ Casing
   ___ Chair rail
   ___ Jambs
   ___ Cased openings
   ___ Window/stool/apron
   ___ Mantle
   ___ Special moldings

☐ Paneling
☐ Interior doors
☐ Shelving/closet rods/shower rods
☐ Mirrors
☐ Hardware
   ___ Privacy locksets
   ___ Passage locksets
   ___ Dummy locksets
   ___ Doorstops
   ___ Bath accessories
   ___ Ironing board
   ___ Dryer vent
   ___ Attic stairs
☐ Plumbing access doors
☐ Stair finishes
   ___ Treads
   ___ Rails
   ___ Risers
   ___ Skirts
   ___ Disappearing stairs
   ___ Spiral stairs
   ___ Handrails and newels
☐ Fasteners
   ___ Nails
   ___ Screws
   ___ Adhesives
☐ Trim Labor

14. Paint and Wallcovering

☐ Paint materials
☐ Wallcovering
☐ Wallcovering labor
☐ Ceramic tile
☐ Paint labor

15. Floor Covering

☐ Carpet and pad
☐ Resilient floors
☐ Wood floors
☐ Hard flooring
   ___ Tile
   ___ Slate/flagstone
   ___ Brick
☐ Installation labor

16. Cabinets

☐ Kitchen
☐ Bath
☐ Special cabinets
☐ Countertops
☐ Backsplash
☐ Cabinet hardware
☐ Cabinet installation labor

17. Appliances

☐ Dishwasher
☐ Range/cook top system
☐ Exhaust fan/range hood
☐ Oven

FIGURE 2.2 *Continued*

17. Appliances (*continued*)

- ☐ Microwave oven
- ☐ Refrigerator
- ☐ Disposal
- ☐ Central vacuum system
- ☐ Ice maker
- ☐ Other

18. Exterior

- ☐ Final grading
- ☐ Landscaping
  - ___ Fill material
  - ___ Topsoil
  - ___ Trees

- ___ Shrubs
- ___ Mulch
- ___ Sod
- ___ Seed/fertilizer
- ☐ Walks, driveway, patios, stoops, and/or steps
  - ___ Formwork
  - ___ Form labor
  - ___ Concrete
  - ___ Concrete labor
  - ___ Base material
  - ___ Spread/compact base material

- ___ Brick/stone/tile
- ___ Asphalt
- ___ Gravel
- ☐ Irrigation/sprinkler system
- ☐ Gutters and downspouts
- ☐ Splashblocks and drainage system
- ☐ Ornamental ironwork
- ☐ Garage doors
- ☐ Fencing
- ☐ Clean
  - ___ Inside
  - ___ Outside

2,250 cubic feet. Divide by 27 cubic feet per cubic yard and you get 83.3 cubic yards to be cut within that grid square.

If a grid square has both cut and fill, draw a line across the square where the cut changes to fill. Then choose two opposite sides of the grid and calculate the area of cut at each end. Next you average the two end areas and multiply by the distance between them to get the volume. For example, say you have a grid as shown in Figure 2.3.

If you consider the top edge, we cut 2 feet at the right end and the cut ends at about 12.5 feet across and becomes fill. The end area of cut at the top edge is, therefore, 2 feet, the height of the triangle × 12.5, the base of the triangle, divided by 2, since the area is a triangle, or 12.5 square feet. At the bottom edge the end area of cut is 3 × 18.75 divided by 2 or 28.125 square feet. The average end area is (12.5 + 28.125) divided by 2 or 20.3 square feet. If you multiply the average end area by 25 feet which is the distance between the ends, you get a total volume of cut of 20.3 × 25 = 508 cubic feet or 18.8 cubic yards. The amount of fill within this grid square is also the average of the two end areas times the distance between. The two end areas of fill are 12.5 square feet for the top and 3.125 square feet for the bottom. The average is 7.8 square feet. If you multiply the average end area by 25, you get 195 cubic feet or 7.2 cubic yards of fill.

Add up the cut yardage from all of your grid squares to get the total cut and all the fill yardage to get the total fill. If the cut and fill do not balance, you will have to haul off or bring in dirt. In many cases you can adjust your final grade and bring the cut and fill into balance.

One of the differences between taking off the quantities of excavation for basements and some deep footings is that you have no detail on the drawings to tell you exactly how much to excavate. In most of these cases you will need to allow for additional working space and side slopes to prevent slippage and caving of the banks. After you have determined the limits of the excavation, estimate the quantities of earthwork the same way you did for the general earthwork. Usually you can determine end areas and, if they are different, average them. Multiplying average end areas by the length will usually give you a reliable number. Of course, this method assumes that the site has no bumps or dips between the ends. If it does, you need to choose shorter lengths to get a better answer.

For example, assume that you have to excavate for a deep, continuous footing. Further, assume that the footing is to be 1.5 feet wide and 3.5 feet underground as shown in Figure 2.4.

In this case the estimator must determine how much working space is needed and what side slopes are needed. Assume that the estimator decides that 1 foot on each side of

FIGURE 2.3 **Earthwork Determination**

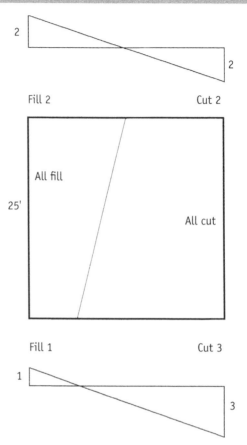

the footing is adequate along with a one-to-one side slope. The end area of this excavation is—

$$\frac{10.5 + 3.5}{2} \times 3.5 = 24.5 \text{ sq. ft.}$$

If the footing is 40-feet long, the excavation is approximately 24.5×40 or 980 cubic feet. This figure may be converted to cubic yards by dividing by 27, which yields 36.3 cubic yards.

Most builders rarely calculate the quantities of backfill required for basements, ditches, and other construction. In these cases, builders rely upon their experience to estimate the time required to backfill and spread out any excess. However, if you need to calculate the backfill quantities, the best way is to determine the volume of the hole and subtract the volume of the construction put in the hole up to grade level.

Most filling operations require compaction. Compaction is done by hand using tamps and vibrating plates or machinery such as sheepsfoot rollers, loaded dump trucks, or bulldozers. If you compact with bulldozers or front-end loaders, be sure to keep the depth of each layer or lift to no more than 6 inches to ensure complete compaction.

You also need to understand the difference between bank cubic yards and loose cubic yards. If you dig a hole that is exactly 100 cubic yards and are hauling the soil away in 10-yard dump trucks, you will need more than 10 loads. The reason is that the soil in the ground is in a compacted state and will have far less air voids than when it is dug up

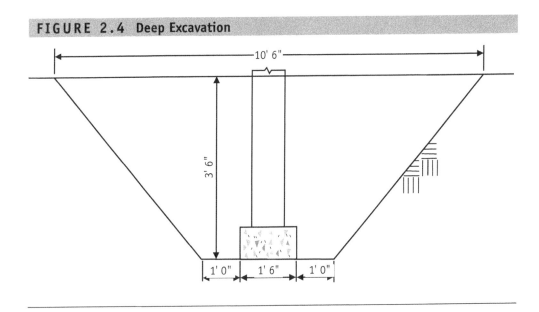

FIGURE 2.4 Deep Excavation

and dumped into a truck. Likewise, if you calculate that you will need a certain quantity of compacted fill, you will need about 25 to 30 percent more soil if you are buying it by the truckload since the volume in the truck is loose measure.

Footings and Slabs

This category includes all the concrete placed near the beginning of the job. Walks, drives, and patios that are placed near the end of the job are in a later category. The concrete item is divided into two categories so that the footings and slab category, which comes early in the job, can be closed out and the appropriate cost-control comparisons made as the project progresses.

Batter Boards

Draw dots and lines on your architectural floor plan to indicate stakes and cross-pieces. Count each one and list them on your worksheets.

Formwork

Estimating formwork for slabs is usually a matter of measuring the perimeter to determine the total length of boards needed. If the ground drops off in certain areas of your house, you will have to determine how many extra boards you will need. You should list the formwork by size and length and the number of boards, for example 30 ÷ 2 × 12 × 12.

When taking off lumber, some builders like to convert the quantities to board feet and estimate the costs based on total board footage. However, the price per board foot will vary depending on the size and length. Many builders simply list the items they need by length and size and price each different piece. By doing it this way, you can easily order from your takeoff sheets. The examples in this book use this method.

You can estimate the number of stakes and braces you need from the perimeter dimension. For example, if you are using 12-foot formboards, you might want a stake every 4 feet. At each stake location you will probably need a brace or kicker and an additional stake for the brace.

Reinforcing Steel

Dowels and anchor bolts may be counted by location. For reinforcing bars, determine how many you will need at each cross-section and multiply by the total length. You will probably buy the bars in 20-foot sections, so allow for an adequate lap. For normal strength concrete and steel, you should lap the bars a distance of 0.4 times the size of the bar (bar number) squared but not less than 12 inches. For example, you should lap a number-six bar $0.4 \times 6 \times 6 = 14.4$ inches. Number-five bars should be lapped 12 inches since $0.4 \times 5 \times 5$ is 10 inches, which is less than the 12-inch minimum. Be sure to look for specially fabricated reinforcing bars such as stirrups and other specialty bends. Welded wire mesh usually comes in 750 square-foot rolls. If you figure 700 square feet of coverage, you will probably have enough to cover your overlap.

Porous Fill

Most specifications call for 4 inches of gravel or porous fill beneath the slab. Taking the total slab area is conservative and will account for waste and variations in the thickness since the porous fill is not placed in footings.

Termite and Soil Treatment

Most pest control companies who pretreat slabs for termite control charge by the square foot. If you need to place termite shields on masonry walls or piers, you will have to count and measure. Radon control techniques vary and should be shown on the plans and specifications.

Vapor Control

Polyethylene film is commonly required under concrete slabs. If you buy enough to cover the entire slab area including the footings, you should have enough to cover the required lap.

Concrete

Concrete for footings is taken off by multiplying the cross-sectional area by the length. For example, if a footing is 16×24 inches, the area is 16×24 divided by 144 or 2.67 square feet. And if the total length of this footing is 240 feet, the volume of concrete needed will be 2.67×240 divided by 27 or 23.7 cubic yards. Table 1 in Appendix B gives footing volumes for various size footings. For example, a footing 16 inches by 24 inches has 0.0988 cubic yards per linear foot. Multiplying 0.0988×240 yields 23.7 cubic yards. Most floor slabs are 4 inches thick. If you divide the area by 81, you will get the required yardage. Table 2 in Appendix B gives the coverage for various thicknesses.

Masonry

To estimate the quantity of brick, measure the square footage of area to be covered by brick and multiply by the appropriate factor (bricks per square foot). Most bricks used in residential construction are standard size ($2^{1}/_{2} \times 8$ inches). The number per square foot will depend on the thickness of the mortar joint. For a $^{3}/_{8}$-inch joint you will need 6.55 brick per square foot; for a $^{1}/_{2}$-inch joint, you will need 6.16 bricks. If the bricks you are using are a different size, you will have to figure the factor. Just add the thickness of mortar joint to each dimension and multiply the length by the height to get the area of one brick with a mortar joint on top and on one end. Divide that area into 144 to get the number of bricks per square foot. Measure the area carefully and then multiply by the number of bricks per square foot for your particular job. Add about 5 percent for waste, or calculate enough to cover small openings.

Concrete block work is done similarly. Just remember that the nominal size of the block includes a $^{3}/_{8}$-inch joint. Therefore an 8×16-inch block, covers 128 square inches including the mortar joint. If you are using 8×16-inch blocks, multiply the area to be done with block by $144 \div (8 \times 16)$ or 1.125 to get the number of blocks needed. On blockwork, 2 to 3 percent waste is usually enough. Stone is usually bought by the square foot or the ton. If you are buying it by the ton, you can ask your supplier how many square feet a ton should cover. Otherwise, you can figure how many tons you need based on the wall thickness. For example, if the stone wall is to be 6 inches thick, each square foot of wall will weigh about 75 pounds. (Stone weighs about the same as concrete, which is 150 pounds per cubic foot on average.) Therefore a ton (2,000 pounds) will cover about 2,000/75 or 27 square feet.

Fireplaces

Face brick or other masonry on the chimney and used as a veneer on the interior is taken off with the exterior brick veneer or other masonry in the masonry category. If the fireplace is made of masonry, the builder will customarily purchase the miscellaneous steel items such as the damper and lintels, the firebrick, the flue liners, and the common brick for the smoke shelf. Usually you will need about 100 firebrick and 1,000 common brick for a standard 48-inch fireplace. The flue liners usually come in 2-foot lengths and start about 5 feet above the hearth. The height of the chimney can be determined from the elevation drawings.

Mortar

Plain mortar mix is used for brick and block; however, some masons prefer a mixture of mortar mix and portland cement or just straight portland cement for stone. Figuring mortar mix or portland cement for stone is difficult because stone can be rough and irregular, and the styles vary. If you do not have a better number, figure that one bag will lay about 10 square feet for stone.

You will need about 7 bags of mortar to lay 1,000 bricks, and 1 bag will lay about 35 regular concrete blocks. A lot of waste occurs with sand, especially if you are building in a neighborhood with children who play in it after work hours. You will probably find that you need 1 cubic yard for every 1,000 bricks. Another way of looking at it is that you will use 1 yard for every 7 bags of mortar. You should check these numbers after a few jobs to see if they are right for your operation.

Metal

A regular-sized box of wall ties (about 500) will cover about 1,700 square feet of wall at a spacing of 16" × 32". Table 8 in Appendix B will help for other spacing. Take off your metal lintels by measuring each opening required and adding about 1 foot to each for support at the ends. Many builders make the mistake of ordering the same-size angle for masonry over a wide garage door as they do for a 36-inch window. If the plans and specifications do not clarify its size, you may be wise to consult an engineer.

Framing

In today's building industry, most builders will furnish all or nearly all the materials required under the framing category. Because this category has so many items, you will probably spend the majority of your estimating time in taking off framing materials. Just follow your checklist to help make sure that you do not leave out anything. You can take off framing lumber any number of ways. You may develop some sort of formula for certain items, and you may count other items. You may convert everything to board feet, or you may choose to list the materials by the size and length, and price each different piece. The only formula used in this book for framing is one for taking off studs. If you buy one for every foot of wall plus one for every corner, tee, door, and window, you should be close. With this one exception, many builders find that if they take off the framing lumber by counting the pieces needed as if they were getting ready to order the lumber, their overall work is reduced, they have fewer pieces left over, and they are short fewer pieces.

Waste is accounted for by ordering the next higher incremental length to do the job. For example, if a floor joist must span 10 feet 6 inches, you would take off a 12-foot piece. When taking off flooring materials, many builders like to make the first and second floors different categories and keep these materials separate. Many published formulas and tables can help you estimate rafter lengths. Several useful tables are reproduced in Appendix B. However, if you count each one, you can fairly easily scale and measure them from the elevation drawings. You may not be able to measure the true length of hips and valleys directly from the plans, but you can measure the horizontal distance from the plan and the vertical distance from the elevation drawings. You can then use the $a^2 + b^2 = c^2$ formula or simply draw the horizontal distance and the vertical distance to scale and measure the length.

Most builders find that the most difficult thing to accurately estimate on the whole job is random-length 2×4s. You will need these pieces for top and bottom plates, bracing, subfacia, lookouts, ledgers, attic bracing, and a host of other things. Formulas do not work well because the requirements for random-length 2×4s vary from house to house. The checklist in Figure 2.2 lists many if not most of the uses for random-length 2×4s. The best way to estimate these is to determine how much you need for every use listed.

Taking off roof decking always involves some waste. At gable ends, some waste will occur at the overhang because of the stagger when your framers make the ends of the plywood break over different rafters. Likewise, at hips and valleys, you will have the angle cut-offs as waste. Some builders will count the pieces of plywood necessary by drawing the layout on the plans while others will figure the finished roof area and add 5 to 10 percent waste.

Nails are usually taken off based on experience for a particular size house. Table 16 in Appendix B gives a way to determine the required nails more closely.

Roofing

Miscellaneous roofing items such as roof edge, valley materials, flashing against walls and chimneys, and ridge vents are usually measured directly from the plans. The roofing material itself is determined from taking the finished roof area and adding a waste factor. Of the different kinds of roofing, asphalt shingles will usually require the least waste—usually 3 to 5 percent. Wood shakes take a little more. The waste for slate roofing depends upon the quality of the slate.

Be sure to take off the starter course which typically runs along all eaves. One bundle of asphalt shingles will usually cover about 80 lineal feet. You must also take off the hip-and-ridge shingles. If hip-and-ridge shingles can be bought separately, ask your building supply dealer how many feet a bundle will cover. If you are using regular three-tab asphalt shingles and making the hip-and-ridge pieces on the job, each bundle will go about 36 feet.

Plumbing, Electrical, Heating, Ventilation, and Air-Conditioning

Trade contractors typically do the work in these categories; however, you might have to furnish some items such as vanity tops, whirlpool tubs, tub and shower surrounds, and lighting fixtures. Occasionally you may want to do an estimate on the entire work to be performed by one of these trade contractors just to make sure that your trade contractor's prices are in line. For plumbing, you can determine a fairly good number for comparison purposes by pricing the fixtures and faucets, including hot water heaters, and adding a per fixture price to cover other materials and labor. To calculate this per fixture price, take three or four jobs for which you have recent bids. Subtract the price of the fixtures and faucets and divide the remainder by the number of fixtures. If the plumbing is concentrated in one area of the house, the per fixture material and labor price should be a little less. Be careful not to rely on this method to price a job unless you have proven to yourself over many examples that doing so will provide a reliable figure.

Electrical trade contractors often price a job by adding the number of "drops" multiplied by a unit price to the number of amps in the service panel multiplied by a dollar amount. A drop is defined as a switch, duplex receptacle, light, or other termination point. If your electrical contractor uses this method or you know of others who do, they might tell you what the two unit prices are. If you cannot find out what they are from anyone in the business, you may be able to determine them from the bids. Usually the per drop cost is two to three times higher than the per amp cost.

HVAC prices may be approximated by the tonnage of air-conditioning or the heating output. However, for different types of systems this price can vary greatly.

Insulation

Insulation quantities are easily determined from the plans. Simply measure the wall, ceiling, and rafter areas to be insulated. Insulation has only a little waste so be sure to deduct the door and window areas.

Drywall

The builder will often have to furnish the drywall board and sometimes the joint compound, tape, screws, and nails. To estimate the quantity of drywall board needed, determine the lineal footage of exterior wall, the lineal footage of interior wall, and the square footage of ceiling area. In determining wall footage, include all openings that are 4-feet wide or smaller and subtract openings larger than 4 feet. Multiply the exterior wall length by the ceiling height. Multiply the interior wall length by twice the ceiling height since interior walls get hung on both sides. By including the smaller openings, 4 feet or less, you will usually have enough to cover waste.

If some rooms have different ceiling heights, go back and add the extra height to the perimeter of the room. By adding the footage required for exterior walls, interior walls, and ceiling, you will come up with a fairly accurate number. If your drywall hangers repeatedly run short, check your waste piles. Drywall hangers and finishers usually charge a unit price (so much per square foot) for every board that is used based on the original size of the board. If they cut 4 square feet off of a 48 square-foot board and just hang 44 square feet, they still get paid based on 48 square feet. This practice provides little incentive for the hanger to utilize the waste effectively. Most will do so, but if you consistently run short, check your hangers.

Interior Trim

This category typically takes more time to estimate than any other except framing. Taking off these quantities involves no magic trick other than making sure that you have included every item. If you utilize your exterior wall and/or interior wall lengths for base, you will usually have base left over. The waste on base is usually only about 2 to 3 percent, and the door openings are usually more than that. Crown molding, on the other hand, is independent of the doors and windows in the usual case. When taking off the casing for windows and doors, you should determine the lengths needed and count the pieces. For all the other items needed, your best bet is to take them off as if you were ordering them by counting and measuring.

Paint and Wall Covering

Painting is commonly done by trade contractors who furnish labor and materials. If you have to furnish the paint, figure about 200 square feet of coverage per gallon per coat. If you have to prime raw wood, the coverage will probably be less depending on the kind of wood and the surface texture. Table 15 in Appendix B will help.

Builders often furnish wall covering but have it installed on a per-roll basis. Most rolls cover about 36 square feet, but some rolls may be different. When estimating the amount of wall to be covered, be sure that you account for matching patterns by allowing a little more waste.

The checklist given earlier has ceramic tile for the walls in this category. You may want to put it in the plumbing section, or if the floors and walls are both ceramic tile, you may want to put it in the floor covering section.

Floor Covering

Most carpet is 12-feet wide. The best way to figure how much carpet to order is to sketch out how the seams will run on the floor plan. Sheet vinyl comes in 6- and 12-foot widths and is estimated similarly. You can estimate other forms of floor covering such as hard tile, slate, and brick by measuring the exact area and adding a 3 to 5 percent waste factor.

Cabinets and Appliances

In most cases, cabinets are build by a cabinet shop or bought from a manufacturer on a modular basis. Appliances are listed individually.

Exterior

Final Grade. The final grade is usually estimated by using your experience with previous jobs. If the job differs substantially from the norm, you may have to calculate the yardage and use a reasonable production rate to determine the time requirements.

Landscaping. Usually landscaping is a trade contract item, but if you opt to do it with your own work force, be sure to figure the preparation of the planting beds and any mulch required in addition to the plant material.

Walks and Drives. Take off all the concrete as in the footings and slabs section. Measure the areas and divide by 81 for 4-inch slabs. Be sure to allow concrete for thickened edges. Measure the length of formwork needed. You can count all other items.

No secrets are involved in taking off the balance of the exterior work such as sprinklers, gutters, splashblocks, and fencing. The last item to estimate is cleaning. Builders often subcontract this work, but you may be able to do it with your own workers. Your experience will be your best guide.

Other Considerations

You should measure each item in the quantity takeoff in a uniform manner. For example, your final quantities should all be order quantities even though you begin with net-in-place quantities, which does not include waste. If you take off the net-in-place quantities and then show your waste allowance, the estimate will be easier to understand when you review it.

Wherever possible, the builder should determine distances from the dimensions given on the drawings because many drawings are not drawn precisely to scale.

Many builders use a highlighter pen to mark items of work and materials as they count and measure them to avoid double-counting.

Avoid using shorthand notes in your takeoff unless you describe what each shorthand note means. Many estimators spend a lot of time estimating all the dimensions to the nearest fraction of an inch while others round off to the nearest foot. Devoting a lot of time to estimating some quantities may not be justifiable. For example, in determining excavation quantities, one scoop of the backhoe can exceed all the fractions used to come up with what the estimator thinks is an accurate number. Of course, items that you count should be counted as precisely as possible. As estimators compare the quantities they have estimated to the quantities actually used and recorded as part of the cost control system, they should get a feel for when extra accuracy is justified and when it is not. In any event an estimator cannot afford to be superficial or careless when preparing an estimate.

3

Cost Division

To be a successful estimator you need a few important qualifications. First, you need to understand basic math because estimating involves counting, adding, subtracting, multiplying, and dividing. Second, you need to know and be able to use basic formulas for areas and volumes. (Appendix A has a chart with commonly used formulas). Third, you need a knowledge of construction work methods and operations and to be able to read plans. As part of this understanding of how things are put into place, you need to be able to distinguish between in-place quantities and purchase quantities. For example, you will probably want to buy more drywall than actually ends up on the wall, but you will probably not want to buy extra doors. You need to be organized in the way you compile your data when performing your estimates. A person can be a mathematical genius who knows everything about building but still not be a good estimator without some sort of organization. Finally, an estimator needs a basic knowledge of local building codes and ordinances because certain jurisdictions may require specific methods or materials that will directly affect the cost of the construction.

Understanding how all the elements of cost are related is of primary importance in organizing an estimate.

Direct Job Costs

Estimators usually begin their cost compilation by determining all the direct costs for a job. The first step in this part of organizing the estimate is identifying each and every item or activity that will have a cost associated with it. Your checklist will help you identify these items. The second element in organizing the estimate of the direct cost is breaking the cost of the work items into four basic categories:

- materials
- labor
- subcontracts
- equipment

A work item may have costs in only one of the categories; for example, the heating, ventilation, and air-conditioning (HVAC) cost may be entirely subcontracted and have no material, labor, or equipment components. An item also may have costs in more than one of the divisions. A concrete slab may have material cost for the concrete, subcontracted costs for the placing and finishing, and equipment costs for a concrete pump. Organizing the estimate of the direct cost work items into these four categories offers excellent review and control capabilities.

To do accurate estimates, you need to completely understand how all the costs associated with a house fit together and relate to each other. In order to understand this relationship, you have to define and expand upon each type of cost.

Material Costs

As mentioned, some of the materials incorporated into a house are supplied by the builder and some are supplied by the trades. Therefore the builder only needs to take off the materials that he or she will furnish.

As the builder performs the material takeoff, he or she goes down each item on the checklist and determines the quantity of materials that must be purchased for that item. The builder makes a list of each of the different items needed and gets the unit price of each one. Sometimes standard prices are used, but you should check the prices for each job unless you have a firm understanding with your material suppliers. Get written quotations from suppliers stipulating terms of payment, taxes, delivery charges, and quantities.

Most of the time suppliers will unload their materials at the site themselves, but if you are buying an item you do not often use from a new supplier, check whether or not the supplier unloads. For example, in some areas, suppliers who sell fieldstone for masonry deliver it on pallets and do not unload it. For some materials, you need to know if the supplier will place the material where you need it. For example, if your drywall supplier will stock the required number of boards on the second floor, you must know how this will be done—up the stairs or through the wall.

If you estimate a custom home with specifically named products and want to substitute another brand, make sure you get approval for your substitution ahead of time. Likewise you should ensure that material substitutions proposed by your trade-contractors are preapproved.

If the owner of a custom home is to provide any materials, make sure that you both understand when and how these materials will be available. You must also know the condition of the material and the exact way that material will tie into materials that you will furnish.

If the specifications are incomplete as to some particular item or material, you may have to incorporate an allowance to cover the cost. This practice is customary when the lighting fixtures or hardware items have not been chosen at the time of the estimate. When entering into a contract that has allowances, be sure to determine whether or not sales tax and delivery charges are included. You should treat allowances in the contract as if they were separate cost-plus contracts. You keep up with the costs; if they exceed the allowance, you get paid the difference; if costs are less than allowances, you refund the difference.

Labor

For estimating and accounting purposes, the term *labor* means direct labor on which you pay social security taxes, workers' compensation insurance, and other payroll taxes and benefits. You do not pay all these labor burden expenses on subcontracted labor. Whether or not a particular worker is paid as direct labor or subcontracted labor is not left up to the builder. Some fairly strict rules dictate what constitutes subcontracted labor. In general, if a worker does not take direct orders from the builder on how to do the work on a minute-by-minute basis but instead is required to perform the task in such a way as to achieve an end result, the worker is a subcontractor. Classifying a worker as a subcontractor who is actually a direct laborer can lead to the imposition of fines and penalties. Check with your accountant or attorney if you have any doubts.

Labor is the most risky cost in construction estimating. The labor cost can vary from house to house for the exact same task depending upon the weather, working conditions, attitude of the workers, and many other factors. Since the cost for labor is a function of how long a task takes and since the productivity of the workers can vary, labor costs are difficult to figure accurately.

One way to figure labor cost is to figure how many worker-hours a task should take. This method is called the productivity or production method. Basically, if you know how many units of work are to be done and how long a worker takes to do a unit of work, you can figure how many worker-hours are required. Productivity rates are best determined from your historic records. You may be able to create these productivity rates from time cards or from actually going to the field and timing the work. For example, assume that on average, counting for set-up time and wrap-up time, a mason and a helper can lay 34 concrete masonry blocks per hour. If you have 1,000 blocks to lay, this work should take about 30 hours. If you do not have historic records, you can find productivity rates in many different estimating guides. [4] In determining the labor cost based on the productivity method, you should consider crew composition and the output of the crew as a unit. In doing so, you must use the composite or average wage rate.

A second method to determine labor cost is the unit cost method. In today's home building market, you can subcontract virtually every task. Most of the tasks that you might consider doing with your own forces can be subcontracted on a unit cost or unit price basis. For example, carpenters usually frame a house by the square foot of covered area, and masons lay bricks based on a price per thousand. You can estimate your labor by the unit cost method and compare it with what a trade contractor would charge you for the same work. When using the unit cost method to estimate labor cost, you must judge how much and when to change those unit costs that may vary from time to time and job to job.

Labor burden or indirect labor costs are those costs that a builder must pay in addition to the base salary. Labor burden includes workers' compensation insurance, the employers' contribution to social security, unemployment insurance, other insurance, and fringe benefits such as health plans, retirement plans, and paid vacations. Labor burden is expressed in terms of a percentage of the payroll and varies from craft to craft. This variance is largely attributed to the different workers' compensation rates for different tasks.

Indirect labor costs usually add from 25 to 35 percent to the payroll costs. That is, if a laborer is paid $10 per hour and the labor burden is 29.5 percent, you would add $2.95 to the base pay and use $12.95 per hour for that worker's rate. You can add these costs to the estimate in various ways. Some builders total all their direct labor costs for the whole project and then add an average labor burden rate. Another method is to burden the labor figure at the category level. The method used in this book is to add the labor burden to the payroll costs for each category. Say, for example, that your total estimated labor costs for framing are $10,000 and the labor burden for carpenters is 31 percent; then you would add 31 percent of $10,000 or $3,100 to the cost of that category. Because the workers' compensation rates vary to a considerable degree from trade to trade, adding the labor burden at the category level will give you a more accurate estimate.

Equipment

The term *equipment* as used in this book means only those items such as bulldozers, scaffolding, and power tools that are used to accomplish the work but are not incorporated into the work. Equipment may be divided into three categories. Each type of equipment is estimated differently.

- Major equipment that is on the job for a short period of time and is utilized most of that time. Such equipment might include earth-moving equipment, rough terrain cranes, and concrete pumps.
- Support equipment that is on the job for long periods of time such as power tools, generators, scaffolding, and compressors.
- Small tools whose cost does not exceed some limiting value, say $300, such as hand tools, wheelbarrows, extension cords, and hoses.

Major Equipment

Major equipment is priced similarly to labor. To determine the price multiply the total number of hours necessary to do the task by the hourly price of the equipment. The total number of hours to do the task may be derived from calculating the amount of the work to be done and dividing by the production rate of the equipment.

Example—If you have to excavate 300 cubic yards and the proposed equipment can dig 30 cubic yards per hour, the task should take 10 hours. Many builders estimate the time required for major equipment based on past experience. If you have used a backhoe to dig footings on 50 houses of the same general size, your experience on a per job basis may be more reliable than performing a detailed estimate of the cubic yards to be excavated and dividing by the production rate.

Equipment may be rented or owned. If the equipment is rented, the cost that you apply to the time required is simply the rental rate. If you own the equipment, you should calculate your cost and apply that to the estimated time as an internal rental rate. The cost of builder-owned equipment is made up of two components—ownership cost and operating cost. Ownership cost includes—

- depreciation over the equipment life in hours
- taxes
- storage
- insurance
- investment cost on the average annual value.

Example—A piece of equipment has a delivered cost of $50,000 and an expected life of 5,000 hours over a 10-year period with no salvage value. The hourly depreciation would be $50,000 ÷ 5,000 hours or $10 per hour.

You can calculate your storage costs, insurance, and taxes, if any, directly from your records. Say for this example that they total $2,000 per year. The hourly cost would be the annual cost divided by the number of hours that you use the piece of equipment per year.

In this example, using the the machine 5,000 hours over a 10-year period means that the annual usage would be 500 hours. The hourly cost of storage, insurance, and taxes equals—

$$\frac{\$2,000}{500} = \$4/hour$$

In *Construction Contracting* (Claugh et al., see Resources), the authors offer the following formula for figuring out the average annual value:

$$A = \frac{C(n+1) + S(n-1)}{2n}$$

A = the average value of the equipment

C = the delivered cost

n = the number of years of useful life

S = the salvage value[5]

In the example cited above, the average annual value would be—

$$\frac{\$50,000(10+1) + 0(10-1)}{2(10)} = \$27,500$$

If interest is 8 percent, this annual cost would be $27,500 × .08 = $2,200. The hourly cost would be $2,200/500 hours = $4.40. If you total all the hourly costs, the ownership cost would be—

Depreciation	$10.00
Storage, insurance, and tax	4.00
Investment cost	4.40
Hourly ownership cost	$18.40

In addition to ownership cost, you will have operating costs, which include such items as—

- fuel
- repairs
- tires
- parts
- oil and lubricants
- filters

These costs may be determined from your historical records based upon use. For this example, you may find that these costs total $8.50 per hour. You may determine your total cost of the equipment by adding the ownership cost to the operating cost, which in this example would be $18.40 + $8.50 or $26.90 per hour. Most builders do not include the cost of the operator in the operating cost. This labor cost is subject to labor burden.

If you hire a trade contractor to bring equipment to your job and operate it, this cost is more properly estimated as a subcontract. If the trade contractor quotes you an hourly price, you must still estimate the time that the work will require.

Support Equipment

Support equipment is on the job for an extended period of time, and it is estimated differently from major equipment that is used for relatively short periods. As you perform your estimates, determine those pieces of equipment above some cost level, usually about $300, that you will need for long periods. Estimate how long you will need the equipment and multiply that time by the appropriate weekly or monthly charge. Some builders fail to include this potential cost because if the equipment is company owned, it is easy to overlook when estimating and doing job cost control.

Mobilization costs of major and support equipment include move-in, erection, dismantling, and move-out costs, and these costs are independent of operating time or production rates. However, you must include these costs in your estimate.

Small Tools

Small tools are those hand tools and other equipment supplied by or paid for by the builder for the use of the builder's direct labor force that cost less than some predetermined amount, say $300. If a tool or piece of equipment costs more than the predetermined amount and is left on the job for an extended period, it is estimated as support equipment. Most builders track the cost of small tools over time and establish a rate as a function of the direct labor cost. For example, if you determine that you are spending $20 for small tools for every $1,000 in direct labor, you should include 2 percent of your direct labor estimate for small tools. Of course, if your workers all supply their own tools, you will have no small tools charges.

Trade Contracts

The cost of trade contractors or is often the largest single element of the cost of a house. You must take several considerations into account regarding trade contractors when you are estimating. First, you must find competent trade contractors and evaluate their bids. Having a stable of competent, reliable trade contractors is essential to a modern builder's success. Many builders like to use the same trade contractors over and over, and many builders like to shop around. One advantage to using the same trade contractors is that you form a working relationship. It is usually easier to get a trade contractor to take care of warranty work if that trade contractor is working on other projects for you. The advantage of shopping around is that you can be assured of competitive prices.

You can find potential trade contractors in several ways. Material suppliers and vendors can often point you in the right direction. Other builders can help. Going into neighborhoods under construction and observing trade contractors' work is a favorite way for many builders.

After you find trade contractors, you must evaluate their bids. If more than one such contractor is bidding on a job, they may not all bid on the same exact scope of work. You must make sure what the trade contractor's price includes in order to make a proper comparison and to ensure that every cost is covered. If the trade contractor does not have workers' compensation insurance, be sure to clarify whether you will deduct the appropriate premium from the trade contractor's price. If not, then you must include that cost in your estimate.

In evaluating trade contractor's bids you must consider not only price but also such factors as schedule and reliability. The lowest trade contractor price is not always the best price especially when a trade contractor is unreliable or does poor work. Builders who base their trade contractor choices strictly on price often end up with more headaches and less profit overall. A reliable trade contractor who will be on the job when promised, do quality work, take care of warranty work, and who is financially stable will allow the builder more time to do more work.

Organize Direct Job Costs

The quantities used to estimate the direct job costs are taken off directly from the plans, and the quality of the materials is determined from the specifications. In order to price these quantities, you need to have many different prices and factors readily available. In particular, you will need to know—

- conversion factors
- waste factors
- sales tax rates
- material unit costs
- labor productivity rates
- labor rates
- labor burden percentages
- equipment productivity rates

These figures and factors need to be listed in a single place such as a a database or price book.

Using the appropriate conversion factors and waste factors, you can determine the order quantities and work quantities. Multiplying the order quantities of materials by their unit prices and adding the sales tax yields the material costs.

From the work quantities you can determine the required labor hours by using the corresponding labor productivity rates. Multiplying these hours by the labor rates and adding the labor burden yields the labor cost.

To calculate the equipment hours multiply work quantities for equipment by the proper equipment productivity rates. Multiplying the hours by the equipment rates yeilds the equipment cost.

The trade contractors will use the plans and specifications to derive their cost.

A flow diagram depicting the determination of the direct job cost is shown in Figure 3.1.

Job Overhead

There are many costs that are charged directly to the job that cannot be charged to a particular work item. These costs are called job overhead, project overhead, or job soft costs. Included in this category are such items as the building permit, utilities consumed at the site, sanitary facilities, sales commissions, and interest on the construction loan.

The only reasonable way to estimate the job overhead for a project is to figure the cost of each and every item covered.

Job overhead can vary greatly depending upon whether the house is a custom or contract job or a speculative one. As a new builder, you may have to depend on a general checklist like the one in Figure 2.2 in Chapter 2. However, after you have tracked the costs on several jobs, you will be able to assemble your own list. When you build in an area new to you, you need to check with the local building official to find out how the local jurisdiction prices permits and how it structures inspection fees. You might also check with the local sewer and water boards to see if they charge tap fees. If an item that is to be charged to job overhead is subject to sales tax such as ice or materials for a job sign, you should take it off as materials when estimating these costs. Take off the superintendent's salary as labor to which you must add labor burden. If an item under the job overhead category is not taxed or burdened, it should be taken off under the trade contract heading.

FIGURE 3.1 Direct Cost

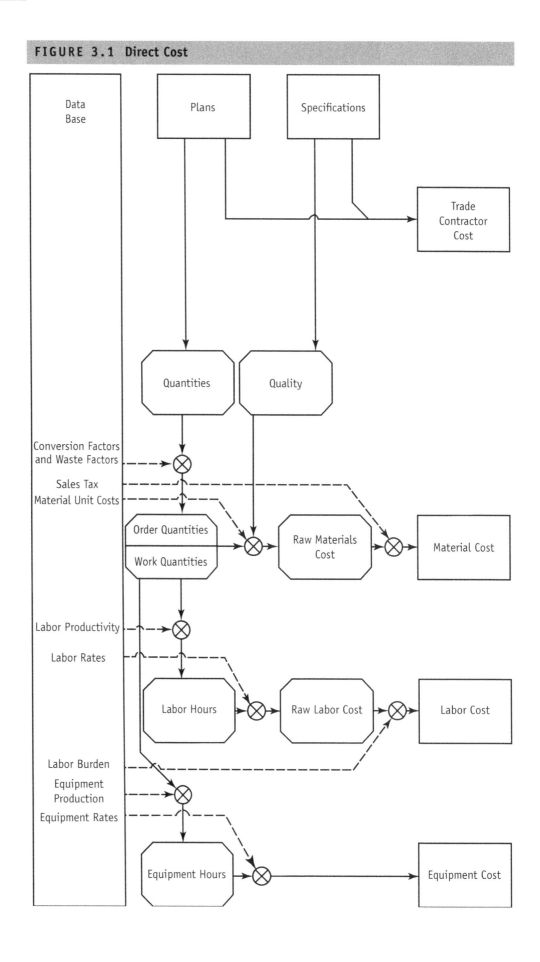

General Overhead

General overhead or home office overhead expenses are those expenses incurred at the main office that cannot readily be charged directly to a job. Such expenses include office rent, executive salaries, office supplies, office furniture, office equipment, telephone, and company vehicles. Even though some builders charge items to the general overhead that could easily be charged to the individual jobs, the better practice is to charge an expense to a job if possible. Because general overhead expenses cannot be charged directly to the job, they must be accounted for in some other way. The way to account for general overhead or indirect expenses is to let each job pay for or absorb its pro rata share based upon what percentage of the total work that job comprises.

In other words, if a job is 10 percent of a builder's work for a given period, that job must absorb 10 percent of the general overhead. In order to estimate how much this amount is, the builder simply divides the company's total general overhead costs for a particular period by the company's direct job costs for that same period. For example, if a company's general overhead for the previous year is $200,000 and the direct job costs were $4,000,000, the general overhead is $200,000 divided by $4,000,000 or 5 percent. After estimating all the direct expenses for a job, add 5 percent to cover the general overhead that a job must absorb. You must figure your own general overhead percentage. Do not rely on this example or what other builders have told you in order to establish your percentage.

Some builders do not include their own salaries in the general overhead but take it out of profit instead. The better way is to figure a reasonable salary for yourself and include it in the general overhead. You should pay yourself an amount comparable to what you would have to pay someone else to do the things that you do. If you have a good year, you can give yourself a bonus. If you have a bad year, you do not have to take your full salary; the company can owe it to you. In this way you recognize the value of your work and account for it properly.

Profit

Profit is not a cost to the builder; it is a cost to the buyer. Like general overhead, profit is added as a percentage of the cost. If you have all of your indirect costs covered in your general overhead, including your own draw or salary, profit is what you have left over after all the expenses are paid.

Several factors determine how much profit you add to a job. As a general rule, a builder should put as much profit on a job that will still allow him or her to get the job because the value of any product is what a buyer is willing to pay. Some builders add a profit and a separate contingency at the end. These two items are really the same thing. In determining your profit for a particular job, you should consider all those items that would make your costs increase. Factors such as who the architect is, who the owner is, the weather, the complexity of the project, and the location should all play a part in determining how much profit to add. Because the profit and the general overhead are usually expressed in terms of percentages, they are usually combined into a single percentage called markup.

Example—Your general overhead is running at 6 percent, and you want to make 12-percent profit, you would multiply the direct costs by 18 percent and add that number to the direct cost to come up with the contract price. Another way is to first add the 6 percent to the general overhead costs and then add the profit based on the total cost. If you do it this way, you are adding profit on the general overhead. For example, if a job has a direct cost of $100,000 and you mark it up with a combined mark-up of 18 percent, the price would be $118,000. However, if you add 6 percent general overhead, the total cost would then be $106,000. And if you add 12 percent of this number, the price would be $118,750.

Relationship of Costs

The builder needs to understand all of the elements of the contract price to understand the relationship between them. Figure 3.2 graphically shows how all the costs are related.

The contract price is made up of the total cost and the profit. The total cost may be broken down into direct cost and general overhead. Together, the general overhead and the profit compose the markup. The direct costs may be subdivided into work item cost, which is the cost of the materials, labor, equipment, and trade contracts, plus the job overhead.

FIGURE 3.2 Cost Relationships

Contract Price					
Total Cost					Profit
Direct Cost				General Overhead	Profit
Work Item Cost			Job Overhead	Markup	
Materials	Labor	Equip-ment	Trade Contracts	Job Overhead	Markup

4

Prepare a
Detailed Estimate

After you study the plans and specifications and perform a site visit, the next step is to take off the quantities of materials that you will need to furnish. After you estimate the quantities, the next step is to price the different elements of the work. The general procedures for establishing a checklist and performing a quantity takeoff were discussed in Chapter 2. The actual pricing of the work is done in a separate process as discussed below.

Pricing Direct Costs

As discussed earlier, the direct cost includes the costs of all the materials, labor, trade contracts, and equipment plus the job overhead functions that make up the job.

Pricing Materials

Your quantity takeoff will give you a list of all the materials that you will need to furnish. These materials typically include items to be purchased from several suppliers such as fill dirt, lumber, concrete, masonry, sand, plumbing fixtures, lighting fixtures, drywall, paint, wall covering, carpet, appliances, and landscape materials. Using your takeoff, make an organized list of all the different kinds of materials you will need. If you are buying a large variety of materials from one supplier, you may wish to give a list of these items to the supplier and get him or her to write in the unit prices. E-mail makes this step especially easy. You may price some items over the phone. Since some material prices can change rather rapidly, the prices you used a month or two earlier may no longer be valid. You cannot afford to take chances in pricing materials. You should go through the procedure of obtaining unit prices on every job unless you have an agreement from your suppliers to hold the prices firm for a specified period.

Pricing Labor

Labor is defined as those workers paid hourly from whom you withhold taxes. As discussed earlier, estimating labor costs may be done in one of two ways: determine (a) how long a task should take and what the pay rate is or (b) what unit price a task should be.

In pricing the labor that you pay for by the hour, you have to take into account the probable working conditions. Worker productivity may vary greatly and generally represents one of the greatest estimating risks that a builder faces. The factors affecting productivity include—

- climatic conditions
- working conditions
- amount of overtime being worked
- morale and attitude
- degree to which different trades are on the job at the same time
- incorrect crew size
- defects in the plans
- delays in material or equipment delivery
- owner occupancy
- disruption of continuity

To price labor by the how-long-it-should-take method, you must have some basis for estimating how many units of work a worker can do in a given time period. For example, from your records you may determine that a laborer can dig a 16- × 24-inch footing at a rate of 20 feet per hour. If you divide the total footage to be dug by 20, you will find how many hours it should take. As discussed in Chapter 3, if you do not have records of your workers' productivity rates, you may use cost data guides.

As mentioned earlier, the second method for determining labor cost applies a rate that is comparable to a trade contractor's price. Say, for example, that you have a firm price from a trade contractor to frame a house for $3 per square foot. If your own labor cannot do the work for less, then you should not do the work with your own forces without some compelling reason. From time to time you may find that all the trade contractors' prices for certain work seem to be too high. If this is the case, you may choose to do that work with hourly workers and hopefully keep the trade contractor's profit for yourself. This decision can be risky if your hourly workers make mistakes because in these cases, the risks associated with having to redo faulty work are yours.

Pricing Equipment

Equipment is priced by the hour, day, or month depending on how long it is needed. If the equipment is rented, the price is known. If you own the equipment, you should charge the project an internal rental rate as outlined in Chapter 3.

Pricing Trade Contracts

These contracts are the surest part if you do them correctly. Many don't go to the trouble to get trade contract prices on every job. Unless you have some arrangement with a trade contractor, you should never guess or "plug in" a price.

Accuracy

What you get out of estimating depends on what you put into it. Just because you sit down and go through the procedure of doing a detailed estimate does not mean that your final number is going to be accurate. The number-one rule to making your estimate more accurate is, "Don't guess when you can make sure." Don't guess at the price for the electrical work when you can get a firm electrical price. Don't guess how much dirt must be moved when you can measure it. Don't guess on the price of materials when you can get a firm quote. Some estimators plug in a number for a line item because it seems like too much trouble to take the time to take it off and price it. The problem is that these estimators get into the habit of plugging in small items and then start plugging in big items. It might be okay to plug $15 for doorstops because an error of 20 percent one way or the other will not matter. But if you plug $1,500 for the doors, the same error might hurt.

The number-one reason for inaccurate or blown estimates is leaving something out. If you have a complete checklist with which you are familiar and you study the plans carefully to find any item of work not on your list, the chances of leaving something out are greatly decreased. Your checklist should be an evolving document. When you find something new that you have never done before, put it on your list. The best way to improve your checklist is to exercise proper cost control. You should post every expense under the appropriate category in the job ledger. If you are incurring some expense that is not in the budget, you will discover it and be able to make the appropriate correction to your estimating procedure.

Contingencies

Some builders like to add contingencies throughout their estimates to help them meet or surpass their budget goals. If you follow this practice, you cannot tell what the job or any category of it should really cost. Furthermore, if the target budget for a category is too high, it is easy to slack off during the work and not worry about costs. The smart way to handle contingencies is to put them in only once—at the end of the job as a part of the profit. In this way your estimate will not contain any "fat" but will represent your best scientific evaluation of the probable cost of the job. Contingencies in an estimate cover potentially costly conditions when you cannot determine their cost at the time of the estimate. If some unforeseeable cost arises, it comes out of profit. Your profit does not have to be the same fixed percentage on every job. You should vary your percent profit for the unique conditions in each job or as discussed in Chapter 3.

Clerical errors are easy to make when dealing with the amount of numbers involved in the average estimate. Using a computer can help cut down on addition, subtraction, multiplication, and division errors, but input errors can plague any estimate. The best way to avoid these types of clerical errors is to be organized and do detailed work. Keep your notes and information in a manner that someone else could finish or work with the estimate, if necessary, with no time lost on questions.

Computerized Versus Manual Estimating

The sample estimate in this chapter is done by hand while the next chapter discusses how estimating is done on a computer. Many builders normally use computers to assist in estimating. Several benefits are gained by using computers instead of manual methods. Computers can perform the necessary computations quickly and without error. In addition, you can easily store and retrieve your productivity and cost data efficiently. Estimators must make important decisions about methods, materials, and equipment during the estimating process. Computers can relieve the estimator of having to compile and manipulate a lot of numbers, thus allowing the estimator to devote his of her energy to decision-making.

Digitizers play an increasingly important role in the actual measurement and counting of quantities. For example, several excellent earthwork programs are on the market that can estimate earthwork quantities with much greater accuracy than can be done by hand. While many builders do not need the degree of sophistication offered by these particular programs, they clearly represent how you can use computers to work with a greater degree of accuracy at phenomenal speeds. The next chapter goes into more detail about the advantages and uses of computers.

A computer is just a tool and is no better than its operator. It cannot replace judgment or experience. Only you can study the plans and determine the best construction methodology and equipment for the job. Builders face unique problems every day, and while computers assist in their solutions, they cannot replace the human estimator. The estimator needs to understand the entire estimating process from start to finish whether a computer is used to assist or not.

Manual Example

Figure 4.1 shows a floor plan, and Figure 4.2 contains the front and rear elevations of a one-and-a-half story detached home. (See Appendixes C and D for larger views.) The construction is slab on grade with stucco, brick, and stone. Specifications are not given, but they would dictate the finishes and materials reflected in the takeoff.

In this sample estimate, all labor is to be subcontracted except for the superintendent.

Figure 4.3 collectively is 23 pages and is the actual quantity takeoff. Note that only materials and work quantities are figured on the quantity or worksheets. The whole purpose of quantity sheets is to make a record of your takeoff process. When you estimate quantities, you should always make detailed notes. The quantity sheets memorialize your effort. If you look on page 46 in Figure 4.3 under the stone section, you will see the notes recording the areas that receive stone with details as to the measurements. This form allows you, or anyone else, to come back to the takeoff at a later date and see how you took the quantities off. Making shorthand notations is human nature, but you need to be sure to put in enough detail and notes so that another person can follow your logic.

Carefully study the takeoff shown in Figure 4.3. Try to understand the way that the quantities were derived. If you have trouble understanding all the numbers, resolve to make your own quantity takeoffs clearer. A blank sample of a quantity takeoff form is shown in Figure 2.1.

The estimate worksheets are used to summarize and price the job. A blank sample is shown in Figure 4.4, and the example project is summarized and priced in Figure 4.5.

The format of the work that is done on the Estimate Worksheets follows the checklist. Note that the checklist given earlier had 18 categories. The first category is land costs. On the estimate, all the costs associated with this category are listed and priced. The form includes a column each for materials, labor, and trade contracts and equipment. The reason that these are in separate columns is that you will need to work on these numbers separately. You will need to add sales tax to the material costs and labor burden to direct labor costs. Typically, no adjustments need to be made to trade contracts or equipment. However, you may have to pay a sales tax on rented equipment

FIGURE 4.1 Floor Plan

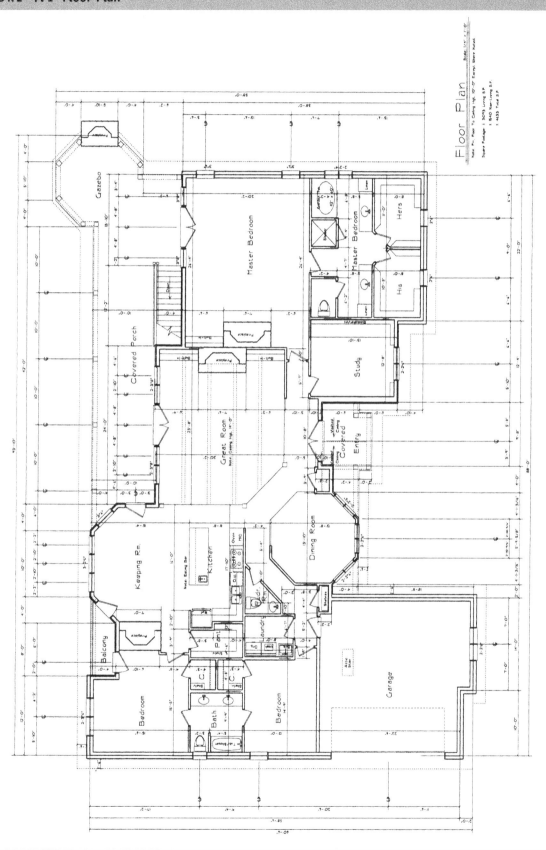

FIGURE 4.2 Front and Rear Elevations

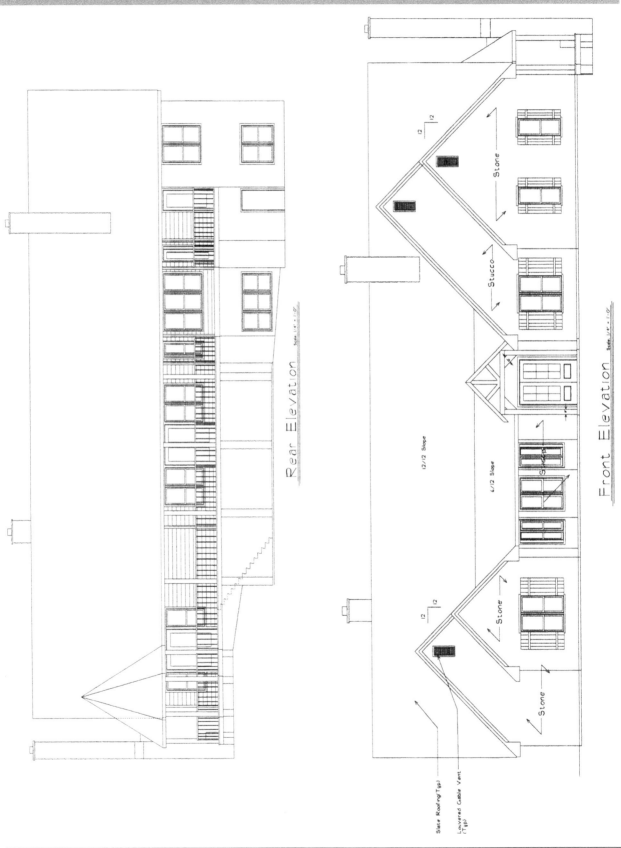

Rear Elevation Scale 1/4" = 1'-0"

Front Elevation Scale 1/4" = 1'-0"

Stone

Stucco

Stone

Stone

12/12 Slope

6/12 Slope

Slate Roofing (Typ)

Louvered Gable Vent (Typ)

FIGURE 4.3 Sample Completed Quantity Takeoff

QUANTITY SHEET

Category: SITE WORK   Estimator: JH   JOB: LOT 2 TURTLE CREEK   Date: 9/28   Page: 1

Description	No.	Dimensions	Subtotal Quantity	Unit	Total Quantity	Unit	Remarks
CUT & FILL							
CUT							
M BATH	①	25' x 27' x 1' /27 CF/CY	25	CY			
MBR	②	20' x 20' x 1' /27 CF/CY	15	CY			
BR3	③	20' x 35' x 1' /27 CF/CY	26	CY			
KEEP RM	④	10' x 11' x 1' /27 CF/CY	4	CY			
					70	CY	CUT
FILL							
GARAGE	①	24' x 24' x 3' /27 CF/CY	64	CY			
DRIVE	②	35' x 50' x 5' /27 CF/CY	324	CY			
					388	CY	FILL
		388 − 70			318	CY	NET FILL (BORROW)

FIGURE 4.3 *Continued*

Category	FTGS & SLABS			QUANTITY SHEET			Date:	9/28
Estimator	JH		JOB: LOT 2 TURTLE CREEK				Page:	2
			Dimensions	Subtotal		Total		Remarks
Description	No.			Quantity	Unit	Quantity	Unit	
BATTERBOARDS			((88+16)+(60+16))×2					
STAKES						20	EACH	2×4×14
CROSSBARS						48	EACH	2×4×12
ANCHOR BOLTS			(88+60)×2 +20+12) ÷4+16			98	EACH	8" ANCHOR BOLTS
			WALLS PIERS					
REBARS			(88×3 +60×2 +20×2+30+20) 2×17			57	EACH	#4 REBARS AT 20 FT
TIEWIRE						1	ROLL	TO TIE REBARS
WIRE MESH			GARAGE 24×24	578	SF			
			DR 10×14	140	SF			
			PORCH 20×6	120	SF			
			STUDY/M BATH 14×12+18×22	564	SF			
			BASEMENT 20×24+20×18	840	SF			
			BACK PATIO 10×26+9×6	296	SF			
				2539	SF			
				÷700		4	ROLLS	6/6 10×10 WWM
POROUS FILL			2539 SF ÷ 81 SF/CY			44	CY	POROUS FILL @ 4" THICK
TERMITE SHIELDS						8	EACH	TERMITE SHIELDS FOR
								16×16 PIERS
VAPOR BARRIER			2539 SF UNDER SLAB +12×20			2	ROLLS	20×100 6 MIL POLY
			+10×18+20×24 +18×32					
			CRAWL SPACE 3835 SF					

FIGURE 4.3 *Continued*

QUANTITY SHEET

Category FTGS & SLABS		Date: 9/28
Estimator JH	JOB: LOT 2 TURTLE CREEK	Page: 3

Description	No.	Dimensions	Subtotal Quantity	Subtotal Unit	Total Quantity	Total Unit	Remarks
CONCRETE							
EXT FTGS		(80+60+10)2x3x1/27	35	CY			
INT FTGS		(18+36+32+26+98) x3x1/27	23	CY			
FIREPLACE FTG		(6x7+12x4+9x5)/27	6	CY			
SPREAD FTGS		(3x3)x22/27	7	CY	70	CY	3000 PSI CONC·FTGS
SLABS		2539/81			31	CY	3000 PSI CONC - SLABS
CMU CELLS		88x3.5/81 +					3000 PSI·CONC -
		140x2/81 +					CMU CELLS
		150x2/81 +					
		17x5x3/81			14	CY	

FIGURE 4.3 *Continued*

QUANTITY SHEET

Category MASONRY   Date: 9/28

Estimator JH   JOB: LOT 2 TURTLE CREEK   Page: 4

Description	No.	Dimensions	Subtotal Quantity	Unit	Total Quantity	Unit	Remarks
FOUNDATION BLOCK							
PERIMETER WALL							
FRONT		2x106 x 1.125	243	EACH			
LEFT		7x66 x1.125	520	EACH			
REAR		7x98x1.125	772	EACH			
RIGHT		2x38x1.125	86	EACH			
INTERIOR WALL							
TRANSITION		3x114 x 1.125	385	EACH			
BASEMENT		10x54 x 1.125	608	EACH			
PIERS		(5+4+4+4+3+3 +2+2+3+32+5 +5+5+5+3+3) x4x1.5	576	EACH			
FIREPLACES		18x15 + 50x5 +14x7	618	EACH			
STAIRWALL		10 x5 x1.125	56	EACH			
HEADER BLOCK		(108 +14 +18+22) 12/16			3164	EACH	8x8x16 CMU-REG
					197	EACH	8x8x16 HEADER CMU
STONE							
FRONT		25x14 +23x14 +8(24+4)	896	SF			
LEFT		2.5x8 + 37x22	834	SF			
REAR		20x12	240	SF			
RIGHT		8x24 +10x22 +10x38	792	SF			
GR FIREPLACE		32x12 + 24x9	600	SF			
KR FIREPLACE		20x10 +12x14	368	SF			
GAZEBO F.P.		12x16 + 12x19	420	SF	4150	SF	STONE

FIGURE 4.3 *Continued*

Category MASONRY			QUANTITY SHEET					
Estimator JH		JOB: LOT 2 TURTLE CREEK					Date: 9/28	
							Page: 5	
Description	No.	Dimensions	Subtotal		Total		Remarks	
			Quantity	Unit	Quantity	Unit		
FIRE PLACES	4							
FIREBRICK					400	EACH	FIREBRICK	
DAMPERS					4	EACH	42" DAMPER	
STEEL ANGLES					4	EACH	54" ANGLE	
COMMON BRICK					4000	EACH	COMMON BRICK	
FLUE LINERS		24x3 +20 = 92 FT			46	EACH	13x13 x 2FT FLUE LNR	
CONC. BLOCKS								
KR. 0-10'		15x6	90	EACH				
GAZ. 0-10'		15x6	90	EACH				
GR/INB/0-10'		15x20	300	EACH				
					480	EACH	8x8x16 CMU	
KR ATTIC		14x12	168	EACH				
GR/INB ATTIC		22x35	770	EACH				
KR CHIMNEY		21x4	84	EACH				
GAZ CHIMNEY		27x4	108	EACH				
GR/INBR CHIMNEY		14x9	126	EACH				
					9521	EACH	8x6x16 CMU	
MORTAR MIX								
BLOCK		3164						
		197						
		480						
		1256						
		5,097 ÷35			146	BAGS	MORTAR MIX	
BRICK		4400 x7/1000			31	BAGS	MORTAR MIX	
STONE		4,150÷10			415	BAGS	PORTLAND CEMENT	
SAND		(146+31 + 415)/7			85	CY	MASONRY SAND	
WALL TIES		4150 ÷ 1700			3	BOXES	WALL TIES	

FIGURE 4.3 *Continued*

QUANTITY SHEET

Category: MASONRY   Date: 9/28

Estimator: JH   Page: 6

JOB: LOT 2 TURTLE CREEK

Description	No.	Dimensions	Subtotal Quantity	Subtotal Unit	Total Quantity	Total Unit	Remarks
METAL LINTELS (STONE)		3 FT III			3	EACH	3 FT METAL LINTEL
		4 FT III			3	EACH	4 FT METAL LINTEL
		6 FT I			1	EACH	6 FT METAL LINTEL
		7 FT II			2	EACH	7 FT METAL LINTEL
METAL LINTELS (CMU)		4 FT II			2	EACH	4 FT CMU LINTEL
		7 FT I			1	EACH	7 FT CMU LINTEL
		9 FT I			1	EACH	9 FT CMU LINTEL
WATERPROOFING							
8' BASEMENT WALL		8X26	208	SF			
10' BASEMENT WALL		10X25	250	SF	458	SF	WATERPROOFING ON CMU
DRAINAGE							
GRAVEL					4	CY	GRAVEL
PIPING		2.5F X 50FT	100	GF	50	LF	3" PERFORATED PIPE
					3	EACH	3" ELBOWS
STUCCO							
FRONT		20X18 = 160					
		12X8 = 96					
		12X10 = 120					
		8X20 = 160			536	SF	STUCCO ON LATH
REAR		24X3+20X5 = 172					
		8X26 + 8X6 = 256					
		10X3+6X9 = 84					
		6X9+5X6 X24 = 210					
		5X3x3x8 128			850	SF	STUCCO ON CMU

FIGURE 4.3 *Continued*

Category FRAMING		QUANTITY SHEET				Date: 9/28
Estimator JH		JOB: LOT 2 TURTLE CREEK				Page: 7

Description	No.	Dimensions	Subtotal Quantity	Subtotal Unit	Total Quantity	Total Unit	Remarks
AREAS							
TOTAL		80¾ x 40	3717	SF			MAIN RECTANGLE
		12 x 1½	6	SF			STONE AT BR 3
		26 x ½	13	SF			STONE AT MBR
		16½ x 25	412.5	SF			GARAGE
		15 x 2	30	SF			GARAGE OFFSET
		10½ x 2	21	SF			FRONT PORCH OFFSET
		6 x 35	210	SF			STUDY-M BATH
		23 x 4½	103.5	SF			M PATH
		14 x 14	196	SF			GAZEBO
		<7-10¾₆ x 7-10¼₆>	<62.4	SF>			GAZEBO
		<4:13/16 x 4-13/16 x ½ x 3>	<25.25	SF>			GAZEBO
		5-9¾"x2	11.6	SF			GAZEBO FIREPLACE
					4,633	SF	TOTAL AREA UNDER ROOF
NON LIVING							
		8 x 4	32	SF			BALCONY
		4 x 4½	8	SF			BALCONY
		10 x 50	500	SF			BACK PORCH
		4 x 4½	8	SF			BACK PORCH
		SEE ABOVE CALCS	120	SF			GAZEBO
		26 x 4	104	SF			BACK PORCH
		22½ x 25	562.5	SF			GARAGE
		<1-10¼₆ x 2¼>	<4.75	SF>			GARAGE
		15 x 2	30	SF			GARAGE
		10 x 6	60	SF			FRONT PORCH
		10½ x 2	21	SF			FRONT PORCH OFFSET
		4'-1 13/16 x 2	8.3	SF			FRONT PORCH
		4'-1 13/16 x 4	16.6	SF			FRONT PORCH
					1,540	SF	TOTAL PORCHES & GARAGE
LIVING		4,633 − 1540			3,093	SF	TOTAL LIVING

FIGURE 4.3 *Continued*

Category FRAMING			QUANTITY SHEET				Date: 9/28
Estimator JH		JOB: LOT 2 TURTLE CREEK					Page: 8
Description	No.	Dimensions	Subtotal		Total		Remarks
			Quantity	Unit	Quantity	Unit	
LINEAL FEET OF WALLS							
EXT. WALLS - 2X4							
		2' GARAGE					
		2' GARAGE					
		10' GARAGE					
		2' GARAGE					
		14' GARAGE					
		22' GARAGE					
		6' DR					
		6' DR					
		6' DR					
			70	LF			EXTERIOR 2x4 WALL
EXT WALLS 2X6							
		12' BACK					
		8' BACK					
		8' BACK					
		6' BACK					
		10' BACK					
		6' BACK					
		6' BACK					
		24' BACK					
		4' BACK					
		26' BACK					
		38' RIGHT SIDE					
		22' FRONT					
		4' FRONT					
		12' FRONT					
		12' FRONT					
		10' FRONT					
		2' FRONT					
		4' FRONT					
		4' FRONT					

FIGURE 4.3 *Continued*

Category FRAMING			QUANTITY SHEET					Date: 9/28
Estimator JH			JOB: LOT 2 TURTLE CREEK					Page: 5
Description	No.	Dimensions	Subtotal Quantity	Subtotal Unit	Quantity	Total Quantity	Total Unit	Remarks
EXT WALLS 2X6 (CON'D)		2' FRONT						EXTERIOR 2X6 WALL
		26' FRONT						
		36' LEFTSIDE	278					
INTERIOR WALLS 2X4		HORIZONTAL (E/W) VERTICAL (N/S)						
		2' BR3	BATH/CL 8'					
		21' BR3/BATH	LAUNDRY/BR2 17'					
		5' CL	BR3 5'					
		39' BR2/CL/KT	PANT/KT 3'					
		2' DR	PANT/KT 5'					
		8' GR	LAUNDRY/½BATH 8'					
		7' LAUNDRY	DR 6'					
		34' MBR/BATH	DR 3'					
		4' HALL/SHELVES	STUDY 2'					
		5' M BATH	M BATH 4'					
		4' M BATH	CL/CL 5'					
		24' MBATH/CL	SHOWER 4'					
		4' ENT/MBR	TUB 4'					
		16' MBR	½ BATH 9'					INTERIOR 2X4 WALL
		14' STUDY	258 LF					
TOTAL EXT WALL		70 + 278				348		TOTAL EXTERIOR WALL
TOTAL INT WALL						258		TOTAL INTERIOR WALL
FLOORS								
SILL PLATES		12+4+8+6+10+6+6+24+9+26+20 +34+2+14+3+6+6+3+26+36+34 = 296 LF						
WOOD BEAMS						19	EACH	2X8X16 PT
						6	EACH	2X12X10
						6	EACH	2X12X16
						4	EACH	2X12X18
						9	EACH	2X12X20

FIGURE 4.3 *Continued*

Category FRAMING		QUANTITY SHEET				Date: 9/28
Estimator JH		JOB: LOT 2 TURTLE CREEK				Page: 10

Description	No.	Dimensions	Subtotal Quantity	Subtotal Unit	Total Quantity	Total Unit	Remarks
JOISTS & BANDS							
BR5		10 / 2×12×10			10	EACH	2×12×10
BR3/BATH		10 / 2×12×14			10	EACH	2×12×14
BR3/CL		3 / 2×12×20			3	EACH	2×12×20
PANTRY		3 / 2×12×14			3	EACH	2×12×14
KEEPING RM		14 / 2×12×20			14	EACH	2×12×20
BR2		16 / 2×12×12			16	EACH	2×12×12
KITCHEN		6 / 2×12×16			6	EACH	2×12×16
KITCHEN		8 / 2×12×10			8	EACH	2×12×10
GREAT RM		14 / 2×12×18			14	EACH	2×12×18
ENTRY		12 / 2×12×8			12	EACH	2×12×8
GREAT RM		2 / 2×12×20			2	EACH	2×12×20
GREAT RM		3 / 2×12×10			3	EACH	2×12×10
MBR		1 / 2×12×12			1	EACH	2×12×12
HKR		17 / 2×12×20			17	EACH	2×12×20
		NOTE: BLOCK 14FT SPAN BR3 & SLAB WALL					
BLOCKING		10 / 2×12×12			10	EACH	2×12×12
BAND		8 / 2×12×12			8	EACH	2×12×12
BALCONY JSC		8 / 2×10×5			4	EACH	2×10×10 PT
BACK PORCH JSTS		17 / 2×10×12			17	EACH	2×10×12 PT
		20 / 2×10×16			20	EACH	2×10×16 PT
		6 / 2×10×20			6	EACH	2×10×20 PT
		3 / 2×10×12			3	EACH	2×10×12 PT
BALCONY BEAM		2 / 2×12×14			2	EACH	2×12×14 PT
		6 / 2×12×14			6	EACH	2×12×14 PT
		5 / 2×12×12			5	EACH	2×12×12 PT
		8 / 2×10×12 BAND			8	EACH	2×10×12 PT
SUBFLOOR		3093 − 631 = 2462 SF ÷ 32			77	SHEETS	3/4"×4×8 PLYWOOD

FIGURE 4.3 *Continued*

Category FRAMING
Estimator JH

QUANTITY SHEET

JOB: LOT 2 TURTLE CREEK   Date: 9/28   Page: 11

Description	No.	Dimensions	Subtotal Quantity	Subtotal Unit	Total Quantity	Total Unit	Remarks
TREATED PLATES							
2x6		30' GARAGE					
		16' ENTRY					
		24' STUDY					
		44' M BATH	114	LF	10	EACH	2x6x12 PT
2x4		52' GARAGE					
		18' DR					
		34' MBR/STUDY/M BATH					
		5' W/C					
		4' SHOWER					
		24' CL					
		18' STUDY/M BATH					
		12' W/C & SHOWER					
		6' CL	173	LF	15	EACH	2x4x8 R PT
UNTREATED PLATE							
2x6		218x3 −114	720	LF	45	EACH	2x6x16 SPRUCE
2x4		(258+70)x3 −173	811	LF	51	EACH	2x4x16 SPRUCE
		1 PER LINEAL FOOT OF WALL + 1 PER DOOR, WINDOW, COR #TEE					
STUDS							
2x6		218 LF WALL					
		17 WINDOWS					
		7 DOORS					
		15 CORNERS					
		15 TEES			332	EACH	2x6x10 STUD

FIGURE 4.3 Continued

Category FRAMING			QUANTITY SHEET				Date: 9/28
Estimator JH			JOB: LOT 2 TURTLE CREEK				Page: 12
		Dimensions	Subtotal		Total		Remarks
Description	No.		Quantity	Unit	Quantity	Unit	
2x4 STUDS			70+258 = 328 LINEAL FEET WALLS				
			4 WINDOWS				
			16 DOORS				
			16 CORNERS				
			19 TEES				
					383	EACH	2x4x10 STUDS
HEADERS	13	2'-6"			7	EACH	2x12x10
	6	3'-0"			3	EACH	2x12x12
	15	3'-6"			8	EACH	2x12x14
	2	4'-6"			2	EACH	2x12x10
	4	5'-6"			4	EACH	2x12x12
	4	6'-6"			4	EACH	2x12x14
	1	9'-6"			2	EACH	2x12x10
GABLE FRAMING		GARAGE 25x6	150	LF			
		M BATH 36x8	272	LF			
		LEFT END 36x9	324	LF			
		RIGHT END 36x9	326	LF			
			1070	LF			
					67	EACH	2x4x16 SPRUCE
BLOCKING		2x6 278 LF			20	EACH	2x6x14
		2x4 258+70 = 328 LF			24	EACH	2x4x14
CORNER BRACING		34 LOCATIONS					
		x1.125 FOR 10FT WALLS			43	EACH	1/2"x4x8 CDX PLYWOOD
SHEATHING		348 LF EXT WALL - 43(4)					
		= 176 LF					
		176÷4x1.125 = 55					
		GABLES: 1070/32=34			89	EACH	1/2"x4x8 EPS SHEATHING

FIGURE 4.3 *Continued*

Category FRAMING							Date: 9/29
Estimator JH		QUANTITY SHEET					
		JOB: LOT 2 TURTLE CREEK					Page: 13

Description	No.	Dimensions	Subtotal Quantity	Unit	Total Quantity	Unit	Remarks
HOUSE WRAP		348 X 10 + 1010	4,550	SF	5	ROLLS	HOUSE WRAP
POSTS		BACK PORCH			13	EACH	8X8X8 WRC
		FRONT PORCH			4	EACH	8X8X10 WRC
BEAMS		BACK PORCH 60+4X6+2+2			3	EACH	8X12X20 WRC
		GAZEBO			2	EACH	8X12X14 WRC
		FRONT PORCH			1	EACH	8X12X8 WRC
		FRONT PORCH			1	EACH	8X12X12 WRC
		FRONT PORCH			1	EACH	8X8X10 WRC
		FRONT PORCH			4	EACH	8X8X8 WRC
		FRONT PORCH RAFTERS			6	EACH	8X8X8 WRC
		GREAT ROOM			4	EACH	8X12X24 WRC
		GREAT ROOM			2	EACH	4X12X24 WRC
		GREAT ROOM			1	EACH	4X12X20 WRC
		GREAT ROOM			1	EACH	4X12X14 WRC
		GREAT ROOM			1	EACH	4X12X10 WRC
GARAGE		GLUE LAM OVER DOOR			1	EACH	3½X13¾X20 GLUE LAM
TRUSSES AT GREAT ROOM		3 14 BETWEEN GR & KEEPING			2	EACH	2X6X10
					1	EACH	2X6X16
					2	EACH	2X6X14
		3 10 BETWEEN GR & ANGLED HALL			2	EACH	2X6X12
					2	EACH	2X6X10
					7	EACH	¾" 4X8X8 PLYWOOD
KNEE WALL		64LF @ 4FT TALL DOUBLE TOP PLATE			8	EACH	2X6X16
		STUDS			28	EACH	2X6X8

FIGURE 4.3 *Continued*

| Category FRAMING | | | | | | | Date: 9/29 |
| Estimator JH | | JOB: LOT 2 TURTLE CREEK | | | | | Page: 14 |

QUANTITY SHEET

Description	No.	Dimensions	Subtotal Quantity	Unit	Total Quantity	Unit	Remarks
CEILING JSTS							
BR 3					15	EACH	2 x 8 x 16
KEEPING RM					13	EACH	2 x 8 x 24
BATH/PANTRY					15	EACH	2 x 8 x 10
BR 2					19	EACH	2 x 8 x 14
DINING RM					14	EACH	2 x 8 x 20
GREAT RM					18	EACH	2 x 8 x 20
STUDY					7	EACH	2 x 8 x 12
STUDY					6	EACH	2 x 8 x 18
MBR					3	EACH	2 x 8 x 14
MBR CLS					19	EACH	2 x 8 x 20
M BATH					6	EACH	2 x 8 x 22
M BATH					16	EACH	2 x 8 x 10
GARAGE					17	EACH	2 x 8 x 24
GARAGE					1	EACH	2 x 8 x 14
BACK PORCH					19	EACH	2 x 8 x 12
BACK PORCH					14	EACH	2 x 8 x 16
GAZEBO					6	EACH	2 x 8 x 20
GAZEBO					4	EACH	2 x 8 x 14
FRONT PORCH					5	EACH	2 x 8 x 12
RAFTERS							
BACK					15	EACH	2 x 10 x 12
BR 3					15	EACH	2 x 10 x 16
BR 3					13	EACH	2 x 10 x 12
KIT					13	EACH	2 x 10 x 16
KIT					18	EACH	2 x 10 x 12
GREAT RM					18	EACH	2 x 10 x 18
GREAT RM					19	EACH	2 x 10 x 12
MBR					19	EACH	2 x 10 x 16

FIGURE 4.3 *Continued*

Category FRAMING			QUANTITY SHEET				Date: 9/29
Estimator JH			JOB: LOT 2 TURTLE CREEK				Page: 15
Description	No.	Dimensions	Subtotal Quantity	Subtotal Unit	Total Quantity	Total Unit	Remarks
RAFTERS							
		FRONT					
		BR2			15	EACH	2 x 10 x 12
		BR2			15	EACH	2 x 10 x 14
		DINING RM			13	EACH	2 x 10 x 12
		DINING RM			13	EACH	2 x 10 x 14
		GREAT RM			18	EACH	2 x 10 x 8
		GREAT RM			18	EACH	2 x 10 x 18
		MBR			19	EACH	2 x 10 x 12
		MBR			19	EACH	2 x 10 x 14
		FRONT PORCH			32	EACH	2 x 10 x 16
		M BATH/STUDY			20	EACH	2 x 10 x 14
		M BATH/STUDY			7	EACH	2 x 10 x 20
		GARAGE			36	EACH	2 x 10 x 20
		GARAGE			5	EACH	2 x 10 x 14
		GAZEBO			24	EACH	2 x 10 x 16
VALLEYS							
		M BATH			4	EACH	2 x 12 x 16
		GARAGE			2	EACH	2 x 12 x 20
RIDGE		88+3+29+30+4+2 =	156	LF	10	EACH	2 x 12 x 16
BRACING							
		BR2, BR3			45	EACH	2 x 6 x 10
					13	EACH	2 x 6 x 8
		KIT/DR			26	EACH	2 x 6 x 10
					26	EACH	2 x 6 x 16
		GREAT ROOM			18	EACH	2 x 6 x 8
					36	EACH	2 x 6 x 10

FIGURE 4.3 Continued

Category FRAMING		QUANTITY SHEET					Date: 9/29
Estimator JH		JOB: LOT 2 TURTLE CREEK					Page: 16
Description	No.	Dimensions	Subtotal Quantity	Subtotal Unit	Total Quantity	Total Unit	Remarks
BRACING		MBR			38	EACH	2 x 6 x 10
					38	EACH	2 x 6 x 18
		MBATH			24	EACH	2 x 6 x 10
		GARAGE			26	EACH	2 x 6 x 8
		GARAGE			26	EACH	2 x 6 x 12
PLYWOOD GUSSETS		GARAGE 5x13 =	65	EACH			
		KIT/DR 6x22	132	EACH			
		MBR 5x19	95	EACH			
			292	EACH			
			x 2	SIDES			
			584	GUSSETS			
			÷ 32	SF/SHEET	19	SHEETS	3/4 x 4 x 8 PLYWOOD
LONGITUDINAL BRACING		24+108+90+56+84 =	412	LF	26	EACH	2 x 6 x 16
STRONG BKKS		2x26 + 4x20 + 6x18 =	276	LF	17	EACH	246 x 16
SUBFASCIA		80+28+18+4+18+22+28=	198	LF	13	EACH	2 x 4 x 16
RAKE SUBFASCIA		28+28+20+20+12+29+29+29+22=	244	LF	16	EACH	2 x 4 x 6
LOOKOUTS		(198+248)x74/16	663	LF	40	EACH	2 x 4 x 12
BAND		SEE SUBFASCIA ABOVE			13	EACH	2 x 4 x 16
ROOF SHEATHING		4633 x 1.414 x 1.1 /32			225	SHEETS	3/4" x 4 x 8 PLYWOOD
FELT		225 x 32 /200			36	ROLLS	30# FELT

FIGURE 4.3 *Continued*

Category FRAMING			QUANTITY SHEET				Date: 9/29	
Estimator JH		JOB: LOT 2 TURTLE CREEK					Page: 17	
Description	No.	Dimensions	Subtotal		Total		Remarks	
			Quantity	Unit	Quantity	Unit		
EXT. MILLWORK								
SIDING		BALCONY 16x10-48	132	SF				
		KEEPING RM 10x10-72	28	SF				
		PORCH 66x10-294	366	SF				
			526	SF x1.71=	900	LF	8¼" HARDIPLANK	
FACIA		80+28+20+30+24					1x8x16 WRC	
		+ 28+20+28+26+18						
		+ 8+20+16+20+22						
		+ 28+28 =	492	LF	28	EACH		
SOFFIT		492x1.5x11/32			23	SHEETS	3/8x4x8 PLYWOOD	
GARAGE CLG		(24x22 + 14x2)x11/32			19	SHEETS	3/8x4x8 PLYWOOD	
PORCH CLG		FRONT 8x16	128	SF	310	LF	2x6 T&G V-EDGE	
		FRONT 4x9	36	SF	94	LF	1x6 T&G V-EDGE	
		BACK 4x2+24x10+26x14						
		+ 6x7 + 6x11	858	SF	2060	LF	1x6 T&G V-EDGE	
		BALCONY 10x4	40	SF	96	LF	1x6 T&G V-EDGE	
FRIEZE			442	LF	28	EACH	1x4x16 WRC	
LOUVERS	5	16 x 48 WOOD LOUVERS			5	EACH	16"x48" WOOD LOUVERS	
SOFFIT VENTS		(80+16+4+4+18+18+22+22)/4			41	EACH	SOFFIT VENTS	
EXTERIOR LOCKSETS					2	EACH	HANDLESET	
					2	EACH	DUMMIES	
					1		LOCKSETS	
					3	EACH	HEAD & FOOT BOLTS	

FIGURE 4.3 Continued

QUANTITY SHEET

Category FRAMING						
Estimator JH		JOB: LOT 2 TURTLE CREEK			Date: 7/29	
					Page: 18	

Description	No.	Dimensions	Subtotal Quantity	Subtotal Unit	Total Quantity	Total Unit	Remarks
WINDOWS	2	3°6² TWIN			2	EACH	3°6² TWIN
	2	2'8'6² TRIPLE			2	EACH	2'8'6² TRIPLE
	4	3°6²			4	EACH	3°6²
	6	3°8°			6	EACH	3°8°
	2	2°6²			2	EACH	2°6²
	1	2'0'6⁴ TWIN			1	EACH	2'0'6² TWIN
	4	2'8'6² TWIN			4	EACH	2'8'6² TWIN
	1	2'0'3²			1	EACH	2'0'3²
	1	2'0'4² TWIN			1	EACH	2'0'4² TWIN
WINDOW WRAP			302	LF	3	ROLLS	PEEL & STICK WINDOW WRAP
TRIM BOARDS		18 CORNERS			18	EACH	1x4x10 WRC
		OVER GARAGE DOOR			1	EACH	1x12x20 WRC
EXT DOORS	3	3°8° FULL VIEW			3	EACH	3°8° FULLVIEW DOOR
	2	3°8° DOUBLE - FULL VIEW			2	EACH	3°8° DBL-FULLVIEW DOOR
	1	3°8° DOUBLE - HALF GLASS (FRONT DOOR)			1	EACH	3°8° DBL FRONT DOOR
	3	3°8° SOLID			3	EACH	3°8° SOLID
STAIRS		STRINGERS			3	EACH	2x12x16 PT
		TREADS 2x14			2	EACH	2x12x14 PT
		RISERS			7	EACH	1x8x6 PT
DECK		(BACK PORCH)					
		SKIRT 12+80+20 =	120	LF -3 SIDES	22	EACH	1x12x16 WRC
		DECKING 10x4+2x4+10x24 +26x6+6x16 =	800	SF	1600	LF	DECK BOARDS 2x6
RAIL		12+46+34+12			164	LF	RAIL- 2x6 S4S WRC
SHUTTERS		2/3°6²			2	EACH	3°6² SHUTTERS
		4/1'6'6²			4	EACH	1'6'6² SHUTTERS
		2/2°6²			2	EACH	2°6² SHUTTERS

FIGURE 4.3 *Continued*

Category ROOFING/etc		QUANTITY SHEET					Date: 9/29
Estimator JH		JOB: LOT 2 TURTLE CREEK					Page: 19
Description	No.	Dimensions	Subtotal Quantity	Subtotal Unit	Total Quantity	Total Unit	Remarks
ROOFING							
ROOF EDGE		80+28+110+120+18+18	174	LF	11	EACH	1'2 X16 WRC
VALLEY					110	LF	COPPER VALLEY
SLATE		46'3 X1,9/4 X1.05 STARTER			70	SQ	SLATE ROOFING
					200	LF	STARTER
FLASHING		20+14+28+14+6			82	LF	COPPER FLASHING
NAILS		70 SQ/16 SQ PER BOX			5	BOXES	COPPER NAILS
PLUMBING							
VANITY TOPS		8+3+5.5+5.5 (5 SINKS)			22	LF	VANITY TOPS
TUBS					1	EACH	32 X60 CAST IRON
					1	EACH	42 X60 FIBERGLASS
W/C'S					3	EACH	WATER CLOSETS
KIT SINK					1	EACH	KITCHEN SINK
SHOWER		14X8+16			128	SF	SHOWER-CERAMIC
SHOWER DOOR					1	EACH	SHOWER DOOR
TUB SURROUND		11X5 +12X2			79	SF	CERAMIC
WATER MAIN					310	LF	1"Ø PVC WATER LINE
ELECTRICAL							
		UNDERGROUND ELECTRICAL TIE IN			310	LF	UNDERGROUND ELECT.

FIGURE 4.3 *Continued*

Category: INTERIOR
Estimator: JH

QUANTITY SHEET

JOB: LOT 2 TURTLE CREEK

Date: 9/29
Page: 20

Description	No.	Dimensions		Subtotal Quantity	Unit	Total Quantity	Unit	Remarks
INSULATION								
ATTIC INSUL		3093+(14+9+18+20)	GREAT RM POP UP			3,154	SF	R-30 CLG
WALL INSUL		(88+52)2+(32+8)10 =		3,200	SF			
		-47x16 -26x16 -22 =		-688	SF	2512	SF	R-19 WALLS
FLOOR INSUL		3093-60-34-14-(4x22)				2469	SF	R-11 FLOOR
DRYWALL		EXT WALL 346x10=		3,480	SF			
		INT WALL 258x10x2=		5,160	SF			
		CLG		3,093	SF			
		GREAT RM 88x4		352	SF			
		GREAT RM CLG (23x20-16)		(442 SF)		11,600	SF	1/2 DRYWALL
INTERIOR TRIM								
BASE		348+258 +258-14x3-16x5				760	LF	BASE
SHOE		760 -(15+12)2 -(16+12)2				650	LF	SHOE
		BR2 BR3						
CROWN		ENTRY/DR		114	LF			
		STUDY		52	LF			
		KIT/KEEPING		84	LF	250	LF	3 PIECE CROWN

FIGURE 4.3 *Continued*

Category INTERIOR TRIM			QUANTITY SHEET				Date: 9/29
Estimator JH		JOB: LOT 2 TURTLE CREEK					Page: 21
		Dimensions	Subtotal		Total		Remarks
Description	No.		Quantity	Unit	Quantity	Unit	
CROWN							
		BR3	62	LF			
		BR2	54	LF			
		MBR	96	LF			
		M BATH	83	LF	300	LF	3 5/8" CROWN
CASING							
DOORS		20x4+34+28+22+20	164	LF			
WINDOWS		26+32+20x3+46x3+22+26x5+22	278	LF	450	LF	3 1/2" CASING
WINDOW STOOL		7x10+4x3+5x2+3x3+6x3	60	LF	60	LF	WINDOW STOOL
MANTLES		ALL STONE EXCEPT MBR			1	EACH	WOOD MANTLE
INTERIOR DOORS	9	2^46^8			9	EACH	2^46^8 DOORS
	5	2^66^8			5	EACH	2^66^8 DOORS
	2	2^86^8			2	EACH	2^86^8 DOORS
SHELVING		BR3	5	LF			
		BR2	5	LF			
		CL	24	LF			
		COATS	4	LF			
		MBR	120	LF			
		PANTRY	112	LF	270	LF	1X12 SHELVING
MIRRORS		BATH 7X3			1	EACH	7X3 MIRROR
		M BATH 2X4½X3			2	EACH	4½X3 MIRRORS
LOCKSETS	6				6	EACH	PRIVACY LOCKSETS
	9				9	EACH	PASSAGE LOCKSETS
DOORSTOPS	23				23	EACH	STRAIGHT DOOR STOPS
	4				4	EACH	HINGE DOORSTOPS

FIGURE 4.3 *Continued*

Category INTERIOR		QUANTITY SHEET					Date: 9/29
Estimator JH		JOB: LOT 2 TURTLE CREEK					Page: 22
			Subtotal		Total		
Description	No.	Dimensions	Quantity	Unit	Quantity	Unit	Remarks
BATH ACCESSORIES	3	PAPER HOLDERS			3	EACH	PAPER HOLDERS
	2	TOWEL RINGS			2	EACH	TOWEL RINGS
	1	TOWEL BAR			1	EACH	TOWEL BAR
DRYER VENT					1	EACH	DRYER VENT W/ 20 FT HOSE
ATTIC STAIRS		DISAPPEARING STAIRS			1	EACH	ATTIC STAIRS–10'
FLOOR COVERING		CARPET					
		BR3	30	YD			
		BR2	25	YD	50	CY	CARPET
		TILE/STONE					
		LAUNDRY	48	SF			
		DINING RM	150	SF			
		STUDY	138	SF			
		M BATH	300	SF			
		BATH	42	SF	688	SF	TILE/STONE FLOOR
		FRONT PORCH	100	SF	100	SF	BLUE STONE
		WOOD					
		ENTRY	272	SF			
		GREAT ROOM	410	SF			
		KIT/KEEPING	408	SF			
		MBR	470	SF			
		½ BATH	25	SF	1535	SF	WOOD FLOOR
COUNTERTOP					60	SF	GRANITE COUNTERTOP
BACKSPLASH					26	SF	BACKSPLASH

FIGURE 4.3 *Continued*

			QUANTITY SHEET				
Category APP/EXT						Date: 9/29	
Estimator JH						Page: 23	
		JOB: LOT 2 TURTLE CREEK					
No.	Description	Dimensions	Subtotal		Total		Remarks
			Quantity	Unit	Quantity	Unit	
	APPLIANCES						
1	WALL OVEN	WALL OVEN			1	EACH	WALL OVEN
1	MICROWAVE OVEN	MICROWAVE OVEN			1	EACH	MICROWAVE OVEN
1	COOKTOP	COOKTOP			1	EACH	COOKTOP
1	DISHWASHER	DISHWASHER			1	EACH	DISHWASHER
1	REFRIGERATOR	REFRIGERATOR			1	EACH	REFRIGERATOR
1	RANGE HOOD	RANGE HOOD			1	EACH	RANGE HOOD
	EXTERIOR						
	DRIVE	CONCRETE	30x30	900 SF	12 CY		3000 PSI CONC
			300x12	3600 SF			
			÷4	900 CF			
			÷27	33 CY			
			x1.9	63 TONS			
		GRAVEL			63 TONS		GRAVEL
	GARAGE DOOR				1	EACH	18x7 GARAGE DOOR
	RETAINING WALL BLKS	3x40x2			240	EACH	RETAINING WALL BLOCKS
	PINESTRAW	100 BALES					

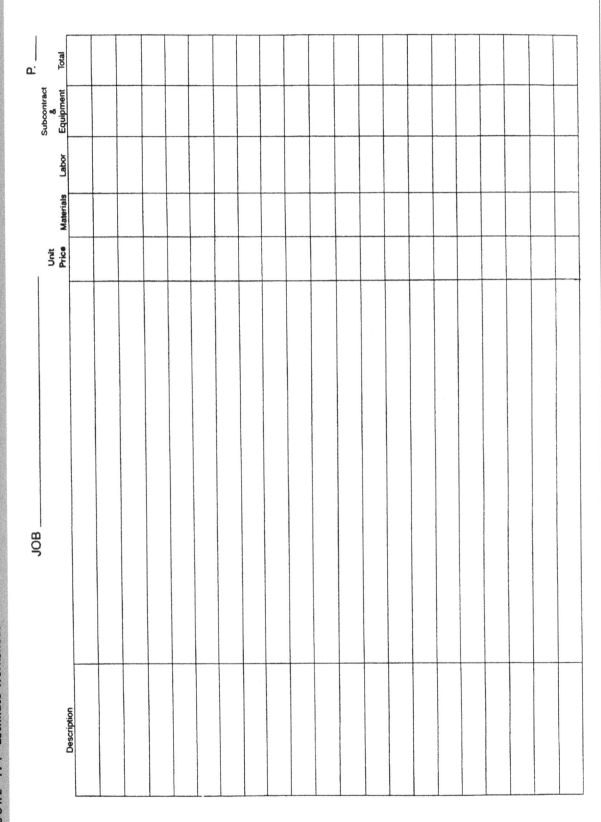

FIGURE 4.4 Estimate Worksheet

FIGURE 4.5 Sample Completed Estimate Worksheet

ESTIMATE WORK SHEET

JOB _LOT 2 TURTLE CREEK_

P. _1_

Description		Unit Price	Materials	Labor	Subcontract & Equipment	Total
LAND COSTS						
LOT					175,000	
SEPTIC SYSTEM					6,000	
					181,000	181,000
JOB OVERHEAD						
PLANS					4,633	
SURVEY LAND & CONSTR. LOAN					500	
CLOSING COSTS						
•appraisal					450	
•discount pts	ONE POINT OR PERCENT. BASED ON A LOAN OF $400,000				400	
•title exam					300	
•doc prep					100	
•recording fee					100	
•abstract					80	
PERMIT	600 + 804 + 753				2,157	
TEMP UTILITIES					200	
BLDRS RISK INS	400,000 AT 6 MONTHS AT $20/MO/100,000	20			480	
GEN. LIABILITY					300	
CONST LOAN INTST	ASSUME BORROWING $80,000/MO FOR 5 MO. = $400,000 1ST MO = 80,000 × 7.5% ÷12 = 500 2ND MO = 160,000 × 7.5% ÷12 = 1,000, ETC 500 + 1,000 + 1,500 + 2,000 + 2,500 = 7,500				7,500	
SALES COMISSION	0.06 × 735,000				44,100	

FIGURE 4.5 *Continued*

ESTIMATE WORK SHEET

JOB __LOT 2 TURTLE CREEK__

P. 2

Description		Unit Price	Materials	Labor	Subcontract & Equipment	Total
SITE EXPENSES						
· Superintendent	$50,000 ANNUAL SALARY ÷ 10 HOUSES PER YEAR			5,000		
· toilet	5 MONTHS	70			350	
LABOR BURDEN	5,000 × 32%			1,600		
				6,600	55,710	62,310
SITE WORK						
SITE CLEARING	$250 MOBILIZATION 20 HOURS AT $80/HR				1,850	
FILL MATERIAL	318 CY	6.00	1,908			
ROUGH GRADING	4 HOURS AT $80/HR				320	
EROSION CONTROL					150	
			1,908		2,320	4,228
FOOTINGS & SLABS						
BATTER BOARDS	20 /2x4x14	4.96	99			
	48 /2x4x12	4.05	194			
ANCHOR BOLTS	98 EACH	0.54	53			
REBARS	57 #4 BARS AT 20 FT	4.62	263			
DIG FOOTINGS & PLACE CONC	(RIKZ CONCRETE)				5,250	
TIEWIRE	1 ROLL	4.38	4			
WIRE MESH	4 ROLLS	67.66	271			
POROUS FILL	44 CY	21	924			
TERMITE SHIELDS	8 AT 24"x24"	4	32			

FIGURE 4.5 Continued

ESTIMATE WORK SHEET

JOB LOT 2 TURTLE CREEK

P. 3

Description		Unit Price	Materials	Labor	Subcontract & Equipment	Total
TERMITE TREATMENT	4633 SF	0.18			834	
VAPOR BARRIER	2 ROLLS 20×100 6MIL POLY	77.29	155			
CONCRETE	115 CY	74	8510			
FINISH LABOR	2,539 SF	0.55			1,396	
CONC PUMP	2 HALF DAYS	800			1,600	
	SUBTOTALS		10,505		9,080	
	SALES TAX AT 8%		840			
			11,345		9,080	20,425
MASONRY						
FOUNDATION BLOCK	3,164 8×8×16 CMU - REGULAR	0.99	3,132		3,164	
	197 8×8×16 CMU - HEADER	1.38	272		197	
STONE	4,150 SF CRAB ORCHARD STONE	5.33 / 5.00	22,120		20,750	
FIREBRICK	400	1.01			400	
DAMPERS	4/ 42" DAMPER	48.17			60	
ANGLES	4/ 54" STEEL ANGLE	9.90	40			
BACKUP BRICK	4000 COMMON BRICK	0.22	880		800	
CONC BLOCKS	480 8×8×16 CMU	0.99	475		480	
	1,256 6×8×16 CMU	0.83	1042		1256	
FLUE LINERS	46/ 13"×13"×24"	14.25	657		184	
SAND	85 CY MASONRY SAND	21	1,701			
MORTAR MIX	177 SACKS MORTAR MIX	5.60	991			

FIGURE 4.5 Continued

ESTIMATE WORK SHEET

JOB LOT 2 TURTLE CREEK

P. 4

Description		Unit Price	Materials	Labor	Subcontract & Equipment	Total
PORTLAND CEMENT	415 BAGS TYPE I PORTLAND CEMENT FOR STONE	7.19	2984			
WALL TIES	3 BOXES	29.07	87			
LINTELS	3 AT 3 FT - STONE	9	27			
	3 AT 4 FT - STONE	12	36			
	1 AT 6 FT - STONE	18	18			
	2 AT 7 FT - STONE	21	42			
	2 AT 4 FT - CMU	13.71	27			
	1 AT 7 FT - CMU	22.50	23			
	1 AT 9 FT - CMU	30.85	31			
WATERPROOFING	458 SF ON CMU	1.50			687	
DRAINAGE GRAVEL	4 CY	21/10	84		40	
PERF PIPE	50 LF	0.86	43			
STUCCO	536 SF ON LATH (ROBLES)	4.25			2278	
	850 SF ON CMU (ROBLES)	2.00			1,700	
			34,029		31,996	
			2,722			
			36,151		31,996	68,747
FRAMING						
SILL PLATES	19 1/2x8x16 PT	14.94	284			
WOOD BEAMS	6 1/2x12x10	12.66	76			
	1/21x12/9	19.96	120			

FIGURE 4.5 *Continued*

ESTIMATE WORK SHEET

JOB __LOT 2 TURTLE CREEK__

P. 5

Description	Unit Price	Materials	Labor	Subcontract & Equipment	Total
WOOD BEAMS					
4/2x12x18	24.12	96			
9/2x12x20	28.70	258			
JOISTS & BANDS					
12/2x12x8	9.80	118			
21/2x12x10	12.66	266			
35/2x12x12	15.46	541			
13/2x12x14	16.20	211			
6/2x12x16	19.96	120			
14/2x12x18	24.12	338			
22/2x12x20	28.70	631			
BALCONY/BACK PORCH BEAMS					
5/2x12x12 PT	18.85	94			
8/2x12x14 PT	19.78	158			
BALCONY/BACK PORCH JOISTS					
4/2x10x10 PT	12.45	50			
25/2x10x12 PT	15.82	396			
20/2x10x16 PT	21.24	425			
6/2x10x20 PT	29.95	180			
TREATED PLATE					
10/2x6x12 PT	7.39	74			
15/2x4x12 PT	5.18	78			
UNTREATED PLATE					
45/2x6x16 SPRUCE	8.82	397			
51/2x4x16 SPRUCE	5.59	285			
STUDS					
332 2x6x10 STUDS	5.82	1,932			
383 2x4x10 STUDS	3.64	1,394			

FIGURE 4.5 Continued

ESTIMATE WORK SHEET

JOB LOT 2 TURTLE CREEK _____   P. 6

Description		Unit Price	Materials	Labor	Subcontract & Equipment	Total
HEADERS	11 / 2x12x10	12.66	139			
	7/2x12x12	15.46	108			
	12/2x12x14	16.20	194			
GABLE FRAMING	67/2x4x16 SPRUCE	5.59	375			
BLOCKING	20/2x6x14	7.23	145			
	24/2x4x14	4.96	119			
CORNER BRACING	43 EACH 1/2"x4x8 CDX PLYWOOD	20.94	900			
SHEATHING	89 EACH 1/2"x4x8 EPS BOARD	8.06	717			
HOUSE WRAP	5 ROLLS	19.00	95			
POSTS	13/ 8x8x8 WRC	117.51	1,528			
	4/8x8x10 WRC	146.99	588			
BEAMS	1/8x12x8	161.00	161			
	1/8x12x12	286.44	286			
	2/8x12x14	334.24	668			
	3/8x12x20	477.48	1,432			
	4/8x12x24	684.48	2,738			
	4/8x8x8	117.51	470			
	1/8x8x10	146.89	147			
	1/4x12x10	101.57	101			
	1/4x12x14	142.20	142			
	1/4x12x14 /1	203.14	203			

FIGURE 4.5 *Continued*

ESTIMATE WORK SHEET

JOB LOT 2 TURTLE CREEK

P. 7

Description		Unit Price	Materials	Labor	Subcontract & Equipment	Total
GARAGE BEAM	2 / 4x12 x24	288.05	576			
TRUSSES	1 / 3⅛ x13⅜ x20' GLUE LAM	138.97	139			
	4 / 2x6x10	4.80	19			
	2 / 2x6x12	6.40	13			
	2 / 2x6x14	7.23	14			
	1 / 2x6x16	8.82	9			
	7 SHEETS 3/4" x4x8 PLYWOOD	28.30	200			
KNEE WAL	8 / 2x6x16	8.82	71			
	28 / 2x6x8	3.54	99			
CEILING JSTS	31 / 2x8x10	6.86	213			
	31 / 2x8x 12	8.23	255			
	27 / 2x8x 14	8.79	237			
	29 / 2x8x 16	11.69	339			
	6 / 2x8x 18	13.10	79			
	57 / 2x8x 20	13.95	795			
	6 / 2x8x 22	25.85	155			
	30 / 2x8x 24	28.20	846			
RAFTERS	112 / 2x10x 12	12.77	1,430			
	72 / 2x10x 14	14.97	1,078			
	103 / 2x10x 16	17.18	1,770			
	36 / 2x10x 18	18.44	663			

FIGURE 4.5 *Continued*

ESTIMATE WORK SHEET

JOB LOT 2 TURTLE CREEK

P. 8

Description		Unit Price	Materials	Labor	Subcontract & Equipment	Total
RAFTERS	45 / 2x10x22	20.45	879			
VALLEYS	4 / 2x12x16	19.96	80			
	2 / 2x12x20	28.70	57			
RIDGE	10 / 2x12x16	19.96	200			
BRACING	57 / 2x6x8	3.54	202			
	169 / 2x6x10	4.80	811			
	26 / 2x6x12	6.40	166			
	26 / 2x6x16	8.82	229			
	38 / 2x6x18	9.11	346			
PLYWOOD GUSSETS	19 SHEETS 3/4"x4x8 PLYWOOD	28.30	538			
LONGITUDINAL BRACING	26 / 2x6x16	8.82	229			
STRONGBACK	17 / 2x6x16	8.82	150			
SUBFACIA	13 / 2x4x16 SPRUCE	5.59	73			
RAKE FACIA	16 / 2x4x16	5.59	89			
LOOKOUTS	40 / 2x4x16	5.59	224			
BAND	13 / 2x4x16	5.59	73			
ROOF SHEATHING	225 SHEETS 7/8"x4x8 PLYWOOD	28.30	6,368			
FELT	36 ROLLS 30# FELT	16.74	603			
SIDING	900 LF 8" HARDIPLANK	0.46	414			
GARAGE CLG	19 SHEETS 7/8x4x8 PLYWOOD	28.30	538			
PORCH CLG	310 LF 2x6 T+G V-EDGE #1 SYP	0.98	304			

FIGURE 4.5 *Continued*

ESTIMATE WORK SHEET

JOB __LOT 2 TURTLE CREEK__

P. 9

Description		Unit Price	Materials	Labor	Subcontract & Equipment	Total
PORCH CLG	2250 LF 1x6 T&G V-EDGE	0.47	1,058			
FRIEZE	28/ 1x4x16 WRC	6.66	186			
LOUVERS	5/ 16x46 WOOD LOUVERS	50	250			
SOFFIT VENTS	41/ 8x16" SOFFIT VENTS	1.78	73			
EXT LOCKSETS	2/ HANDLE SETS	150	300			
	2/ DUMMIES	30	60			
	7/ LOCKSETS	30	210			
	3/ HEAD & FOOTBOLTS	1	1			
WINDOWS	2/ 3'6" TWIN	909.00	818			
	2/ 2/ 2'8'6" TRIPLE	1356.00	2712			
	4/ 3'6"	464.25	1,857			
	6/ 3'8"	443.67	2662			
	2/ 2'6"	378.15	758			
	1/ 2'6" TWIN	741.00	741			
	4/ 2'8'6" TWIN	852.75	3,411			
	1/ 2'0'3"	242.50	248			
	1/ 2'0'4" TWIN	572.25	572			
WINDOW WRAP	3 ROLLS PEEL & STICK 6" WINDOW WRAP	19.00	57			
TRIM BOARDS	18/ 1x4x10 WRC	4.16	75			
	1/ 1x12x20 WRC	36.44	36			
SUBFLOOR	77 SHEETS 3/4"x4x8 PLYWOOD	28.30	2,179			

FIGURE 4.5 Continued

ESTIMATE WORK SHEET

JOB __LOT TURTLE CREEK__   P. __10__

Description		Unit Price	Materials	Labor	Subcontract & Equipment	Total
EXTERIOR DOORS	3/3°8° FULL VIEW	443⁶⁷	2,662			
	2/3°8° DOUBLE - FULL VIEW	819³²	1,639			
	1/3°8° DOUBLE - HALF GLASS - FRONT DOOR	2512⁷⁵	2,513			
	3/3°8° SOLID - STEEL & 9/c JAMB	333⁶³	1,001			
DECK	22/ 1X12X16 WRC	26⁵⁰	583			
	1600 LF DECK BOARDS	2.25	3,600			
RAIL	104 LF WROUGHT IRON	15.42	1,604			
	10 LF 2X6 WRC S4S	1.44	150			
SHUTTERS	2/ 3°6² SHUTTERS	21⁰⁰	42			
	4/ 16² SHUTTERS	12⁰⁰	48			
	2/ 20°6² SHUTTERS	9¹⁰⁰	32			
FRAMING LABOR	4,633 SF (SIMPKINS)	6⁰⁰			27,198	
	SUBTOTALS		75,579		27,198	
	SALES TAX AT 8%		5882			
			79401		27,198	107,199
ROOFING						
ROOF EDGE	11/ 1X2X16 WRC	3.33	37			
VALLEY	110 LF	7.00	770			
SLATE	70 SQ	440	30,800			
	200 LF STARTER	2.66	532			
FLASHING	82 LF COPPER	6.00	462			

FIGURE 4.5 *Continued*

ESTIMATE WORK SHEET

JOB LOT 2 TURTLE CREEK

P. 11

Description		Unit Price	Materials	Labor	Subcontract & Equipment	Total
NAILS	5 BOXES 1¾" COPPER NAILS	162.50	813			
LABOR	705@ (ROBLES)	200			14,000	
	SUBTOTALS (SLATE PRICE INCLUDES TAX)		53,944			
	SALES TAX AT 8%		711			
			33,655		14,000	47,655
PLUMBING						
BASIC SUBCONTRACT PACKAGE	(SHEPHERD PLUMBING)				12,100	
VANITY TOPS	22LF W/ 6 SINKS				1,160	
TUBS	1/32x60 CAST IRON		435			
	1/42x60 FIBERGLASS		265			
TOILETS	3 WC's		390			
KIT SINK	1 KIT SINK + ISLAND SINK		240			
SHOWER	128 SF CERAMIC				640	
SHOWER DOOR		149.27	149			
TUB SURROUND	79 SF CERAMIC				395	
FAUCETS					180	
WATER TIE IN	310 LF				175	
	SUBTOTALS		1,479		15,250	
	SALES TAX AT 8%		118			
			1,597		15,250	16,847

FIGURE 4.5 Continued

ESTIMATE WORK SHEET

JOB Lot 2 Turtle Creek

P. 12

Description	Unit Price	Materials	Labor	Subcontract & Equipment	Total
ELECTRICAL					
BASIC SUBCONTRACT PACKAGE (HAWKINS)				12,150	
LIGHT FIXTURES		ALLOW 4,000		430	
TELEPHONE & CABLE PREWIRE				1,550	
ELECTRICAL TIE IN UNDERGROUND ELECTRICAL TIE IN 310LF				14,130	
SUBTOTALS		4,000		14,130	
SALES TAX @ 8%		320			
		4,320		14,130	18,450
HVAC					
BASIC SUBCONTRACT PACKAGE (McCLUNG)				10,000	
				10,000	16,000
INSULATION					
ATTIC 3,154 SF R-30	.55			1,735	
WALLS 2,512 SF R-19	.36			904	
FLOOR 2,469 SF R-11	.24			593	
SUBTOTALS				3,232	
				3,232	3,232
DRYWALL					
1/2" REGULAR 11,600 SF	28.80	3,345		6,380	
SALES TAX AT 8%		268			
		3,613		6,380	9,993

FIGURE 4.5 *Continued*

ESTIMATE WORK SHEET

JOB LOT 2 TURTLE CREEK

P. 13

Description		Unit Price	Materials	Labor	Subcontract & Equipment	Total
INTERIOR TRIM						
BASE	760 LF	56¢	426			
SHOE	650 LF	17¢	111			
CROWN	550 LF	46¢	253			
ENHANCER	500 LF	71¢	355			
CASING	450 LF	87¢	392			
WINDOW STOOL	60 LF	1.01	61			
MANTLE		200.00	200			
INTERIOR DOORS	9/2⁴68 DOOR UNIT	159.75	1,438			
	5/2⁶68 DOOR UNIT	159.75	799			
	2/2⁸68 DOOR UNIT	164.84	330			
SHELVING	270 LF 1x12	1.39	375			
MIRRORS	1/1x3, 2/4½x3	ALLOW 400	400			
LOCKSETS	6 PRIVACY	20	120			
	9 PASSAGE	18	162			
DOORSTOPS	27	0.54	15			
BATH ACCESSORIES	3 PAPER HOLDERS	6	18			
	2 TOWEL RINGS	11	22			
	1 TOWEL BAR	19.92	20			
DRYER VENT		8.36	8			
ATTIC STAIR	10'STAIR 25½x54"	150.¹²	150			

FIGURE 4.5 *Continued*

ESTIMATE WORK SHEET

JOB LOT 2 TURTLE CREEK

P. 14

Description	Unit Price	Materials	Labor	Subcontract & Equipment	Total
SUBTOTALS		6,655		6,198	
SALES TAX AT 8%		452			
		6,107		6,198	12,305
PAINT AND WALLCOVERING					
PAINT (MARTINEZ)				9,000	
				9,000	9,000
FLOOR COVERING					
CARPET 60 YD	36			1,800	
TILE/STONE 688 SF	5			3,440	
BLUE STONE 100 SF	6.50			650	
WOOD 1535 SF	6.50			9,976	
				15,866	15,866
CABINETS					
CABINETS (COLEMAN)				14,260	
KIT TOPS 60+26=86	65			5,540	
				19,850	19,850
APPLIANCES					
WALL OVEN	1076	1076			
MICROWAVE	617	617			
COOK TOP	2511	2511			
DISHWASHER	924	924			

FIGURE 4.5 *Continued*

ESTIMATE WORK SHEET

JOB _LOT 2 TURTLE CREEK_ P. _15_

Description	Unit Price	Materials	Labor	Subcontract & Equipment	Total
REFRIGERATOR	1,327	1,327			
RANGE HOOD	904 / 904	904			
SUBTOTAL		6,000			
SALES TAX AT 8%		480			
		6,480			6,480
EXTERIOR					
FINAL GRADE (CRITTENDEN)				1,000	
WALK 320 SF (RIKZ)	6			1,920	
DRIVE 12 CY & 900 SF (RIKZ)	74 / 554	888		495	
GRAVEL 63 TONS	13	819			
GARAGE DOOR				480	
RETAINING WALL BLOCKS 240	5 / 2	1,200			
PINESTRAW 100 BALES	3 / 1	300		100	
CLEAN (J&J)				1,000	
TRASH REMOVAL (TWIN CITY)				1,000	
SUBTOTALS		3,207		5,995	
SALES TAX AT 8%		257			
		3,464		5,995	9,459

FIGURE 4.5 *Continued*

ESTIMATE WORK SHEET

JOB ___LOT 2 TURTLE CREEK___

P. 16

Description		Unit Price	Materials	Labor	Subcontract & Equipment	Total
SUBTOTAL	ALL CATAGORIES		188,641	6,600	427,805	625,046
GENERAL OVERHEAD	623,046 × 7.9%					49,220
PROFIT	623,046 × 10%					62,305
SALES PRICE						734,571

SQUARE FOOT CHECK:

$$\text{SALES PRICE} \div \text{LIVING AREA} = \frac{\$734,571}{3,099} = \$237.03/\text{SF}$$

$$\text{COST} \div \text{MODIFIED LIVING AREA} = \frac{448,046}{3,869} = \$115.80/\text{SF}$$

COST = NET COST − LAND COST = $623,046 − $175,000 = $448,046

MODIFIED LIVING AREA = LIVING AREA + ½ NON LIVING AREA

$$3,099 + \frac{1,540}{2} = 3,869 \text{ SF}$$

from an equipment rental company in some categories. Also, you may have to pay workers' compensation insurance premiums for trade contractors who are not covered by their own policies.

The job overhead category deserves special note. You will see that all the costs in this category except for the superintendent's salary are listed in the trade contract and equipment column since none of these charges are subject to sales tax or labor burden.

Some estimators wait until they get to the end of the estimate to total all the material costs and direct labor costs. Then they add sales tax and labor burden. In the example given in this book, the labor burden and sales tax are added at the end of each category. Either way is acceptable. However, if the numbers for each category are taxed and burdened at the category level, comparisons for cost control purposes are easier because the numbers posted in your job ledgers are from bills that have the sales tax included.

For example, you can see that the masonry materials total $34,029, and at an 8 percent tax rate, the sales tax is $2,722. Adding the tax brings the total for materials to $36,751. When the various bills for the masonry come in, they will include sales tax, and you can make a direct comparison.

You will note that at the end of each category, the total cost for that category is listed. The totals for all the categories are added to get the project's direct cost— $623,046 in this example. General overhead and profit are added. You should base yours on your calculated general overhead and how much profit you believe your job is worth. Sometimes builders combine general overhead and profit into a single markup, and sometimes they calculate them separately as in this example. Study the Estimate Worksheets carefully and try to get a sense of the organization of the numbers.

The prices used in the example were reasonable prices on the day and at the place the estimate is done. There is no shortcut for doing a detailed estimate. It is essential that you determine the prices for your jobs from your material suppliers and trade contractors.

Comparisons

In the example, two square-foot cost calculations were made after the estimate was completed. The first is the cost per square foot of living area, which in this case is $237.03 per square foot. As discussed earlier, the sales price and the living area are readily available; real estate agents and buyers often use them for comparison purposes. The second square-foot cost is the cost of the construction less the land costs, $623,046 – $175,000 = $448,046. This total was divided by a modified or effective area, which in this case is the living area plus half the nonliving area: $448,046 ÷ 3,869 square feet = $115.80 per square foot. You would compare this cost per square foot to numbers from your historic database on completed houses.

You also may find the following comparisons beneficial. The cost per square foot of the modified area of the house itself does not include job overhead expenses associated with speculative building. In this example, job overhead totaled $62,310. Of that number, the permit ($2,157), temporary utilities ($200), insurance ($780), and site expenses ($6,950) would be expected if the home were a custom or contract job.

Therefore, speculation expenses totaled $52,223. If this house were sold prior to construction with the owner making progress payments, it would cost $448,046 minus $52,223, or $395,823. The cost per square foot of modified area to build the house would be—

$$\frac{\$395,823}{\$3,869 \text{ sq. ft.}} = \$102.31$$

The land cost as a percentage of the sales price is—

$$\frac{\$175,000}{\$734,571} = 23.8\%$$

The cost per square foot of total area of footing and slabs is—

$$\frac{\$20,425}{\$4,633 \text{ sq. ft.}} = \$4.41$$

The total framing cost per square foot of the total area is—

$$\frac{\$107,199}{4,633 \text{ sq. ft.}} = \$23.14$$

The plumbing price per fixture is—

$$\frac{\$16,847}{15} = \$1,123.13$$

The interior trim cost per square foot of living area is—

$$\frac{\$12,305}{3,099 \text{ sq. ft.}} = \$3.97$$

Once these kinds of costs are determined, they can be compared to actual costs from completed jobs to see if they are in line. Sometimes you can discover a huge mistake in your estimate by making these comparisons.

5

Computerized Estimating

Advantages

Once you learn how to estimate, of course, you will want to do your estimating on a computer, which will greatly improve accuracy, speed, and quality. When you manually estimate a job, you measure and count items on the plans, then list your findings and add them according to their categories. You summarize similar quantities and multiply them by appropriate factors and costs. You make conversions, apply productivity rates, and total up all the costs. For a normal house this process can involve thousands of operations all of which take time and are subject to a variety of errors. Automating these tasks dramatically reduces the time required to do calculations and virtually eliminates computational errors.

Computerized estimating utilizes a custom-made database from which you can draw information as required. You can store prices, conversion factors, and productivity rates pertinent to your operation, modifiy them as required, and use them with speed, accuracy, and ease. Using a computerized estimating system allows more time for quality review of the estimate, especially when it has last-minute changes in the price or in scope of the work.

This chapter offers a brief overview of computerized estimating for home builders. You can discuss your individual company's needs with numerous consultants. Likewise, many of the various software vendors will be glad to provide information to help you set up your own system. In *Estimating With Microsoft® Excel, 2nd Edition*, author Jay Christofferson, a recognized expert in computerized estimating, teaches you how to build a fully customized estimating program that addresses your specific estimating needs (see Resources). Christofferson has also developed a software program (EstimatorPRO™ 5.1, see Resources) to help residential contractors and remodelers can create fast and accurate estimates. In trying to choose what system and what features are best for you, you need to understand some of the benefits of computerized estimating.

Built-in Checklists

Many builders follow a checklist that they continuously update to help them identify and consider all of the work items needed to build a project. You can establish your own computerized checklist to save time in doing the estimate. It also is easily updated. Because many builders are familiar with the items on their checklists, when they study the plans in preparation for beginning an estimate, they should be able to spot work items that are not on their lists. You can easily insert these items into your computerized checklist for inclusion in the takeoff.

Accuracy

Despite ordinary care, every estimator can and does make mistakes such as mathematical errors, transposition of numbers, and misplaced decimal places. You can't get around the fact that if you put garbage in, you get garbage out; however, with a computerized system you can make sure that the math is correct. If you are not saddled with the task of performing calculation after calculation, you can focus on the input and quality of the information being processed.

Standardization

A key factor in producing meaningful, accurate estimates is the organization—how the various costs fit together. Organization helps to make sure that all costs are considered and that nothing is left out. If your estimates are organized in some standard format, they are easier for others to review and check. A computerized estimating system, by its nature, compels your estimates to be standardized. It also offers the added bonus of providing a basis for comparing the estimated and distinguishable parts to previous work. You can easily check square-foot costs, unit costs, and make other meaningful comparisons as outlined in Chapter 4.

Alternative Scenarios

A computerized estimating system allows you to do what-if analyses to see the impact of various alternatives. You can quickly change unit prices and dollar amounts to determine the effect of changing certain materials. You can change productivity rates or quantities and add or subtract major design alternatives. To manipulate these numbers manually would be time-consuming and subject to mathematical error. Being able to consider various design and cost scenarios quickly and easily will help you and your clients to make the best overall decisions with regard to the final design. Your customers will benefit from your ability to help them achieve the best quality and value for their money.

Report and Document Preparation

Your estimate becomes the basis for your cost control system in that the estimated cost for each category of work becomes the budget for that category. A computerized estimating system creates an electronic budget library that you can organize in the best way to suit your needs. The computer can take the numbers used in the estimate to help you prepare a wide variety of reports and other documents. You can prepare and print such items as request for quote sheets for trade contractors, purchase orders for your suppliers, and summaries for your bank. Home building is a competitive business with buyers demanding quality more than ever before. The value of your work is enhanced not only by the quality of your homes but also by the way you run your business. If you can present a professional image through the use of orderly, well-presented documents, you will be ahead of the competition in getting work and making a profit.

Select a Computer Estimating System

Basic Elements

Most estimating systems have two main components: a spreadsheet and a database. The spreadsheet is a form on which the specific information for a project is listed. The sample takeoff (quantity sheets) and the estimate worksheets shown in Chapter 4 are examples of hardcopy spreadsheets. You store the information in the database that you need to perform the estimate. This information includes such items as conversion factors, waste factors, material costs, labor costs, labor productivity, equipment productivity, and equipment rates. One benefit of storing quantities and database information electronically is that if you want to build the same house at some future time, the computer can use your stored takeoff in conjunction with the database information appropriate at that time to give you available up-to-date cost options. The Estimator PRO™5.1 software includes all of the forementioned capabilities (see Resources).

You can choose from many computerized estimating systems. You probably will need many of the features that various software companies offer, but you will not need others. Even though you may not intend to use other computer-based programs such as job costing, accounting, purchasing, or scheduling at first, you would be wise to choose a system to which you can add those options later.

In determining what system to buy initially, keep in mind that hardware and software are only a part of the cost you will incur. You also should consider adding options, training and support, and maintenance costs.

Gather Information

A good way to identify products that you might want to review is to talk to builders who use a computer for estimating. Try to find out the strengths and weaknesses of their systems from their perspectives. Collect and review magazine articles and marketing literature on systems that offer features you need. You can call many of the software companies directly and request information and literature.

If you attend building trade shows, you will almost certainly find exhibitions with computer estimating systems and even seminars on the products they offer. Local representatives of software companies will often be glad to demonstrate their latest products. Of course, they would like for you to buy the most expensive system they offer, so be sure to have what you need in mind ahead of time.

Implement Your System

Get Started

Once you invest in a software system and in the hardware to run it, you must devise a way to phase in the new operation while maintaining your old system. The whole idea of bringing in a new system is to eventually phase out the old one even though you may incorporate parts and concepts of it into your new system. Some builders purchase expensive hardware and software and fail to implement them. Change is difficult, and when people have deadlines to meet, they tend to revert back to old systems and put off using the new one until the so-called right job comes along. Builders who do not use computers for other tasks may tend to postpone learning how to use a new system under the day-to-day pressures of other more urgent activities. Computers, like other tools, require some getting used to. Knowledge of computers is an asset, but it is an asset that you can acquire if you don't already have it. Understanding the construction process and the fundamentals of estimating is far more important.

Consultants

If neither you nor anyone on your staff has the skills and training to implement a computer estimating system, consider hiring a consultant to help you get started. A recommended consultant who specializes in this work and who has experience implementing systems for home builders can save you countless hours of unproductive time by providing valuable expertise.

Consultants may be found in many different ways. Computer software companies keep lists of authorized consultants who stay abreast of the latest advancements. Other builders can often point you to the right person. When considering a consultant, ask for and check references. Once you select a consultant, be sure you have a clear understanding and written agreement on exactly what services are to be provided and how you will pay for them.

Establish and Maintain a Database

You will need time and effort to get your computer estimating system fully operational. You must be willing to devote the time, energy, and money to make it work. Perhaps the most important step in implementing your computer estimating system is establishing your estimating database. This database lists all the numbers, formulas, and factors that you repeatedly use in estimating the price of a house. Examples of some of these numbers are as follows—

- On page 5 of Figure 4.3, Sample Completed Quantity Takeoff, you will need 7 bags of mortar per 1,000 bricks.
- Further on that page, you need to buy 1 cubic yard of sand for every 1,000 bricks.
- On page 11 of Figure 4.3, Sample Completed Quantity Takeoff, the formula for studs is 1 per lineal foot of wall plus 1 for every window, door, corner, and tee.
- Figure 4.5, Sample Completed Estimate Worksheet, lists the unit price for each item of material to be purchased for the sample house.

You may build your database from scratch, in which case you need to organize it in a some logical manner. One way is to list your data according to the way the items are listed on your checklist, which should also be the way they appear in your job cost control ledgers. Some software companies provide ready-made databases that you can easily modify. The prebuilt databases typically include material items, production rates, and other information that you can modify to fit your particular needs. If you choose a ready-made database, be sure to check all the formulas and factors to ensure that they conform to your particular way of building and your experience.

Your database will need regular updating. Don't get so busy that you leave out-of-date prices in your program.

Utilize Your Spreadsheet

When using a computerized estimating system, you use the electronic spreadsheet in the same way that you use the quantity sheets and estimate worksheets shown in

Chapter 4 (Figure 4.3 and Figure 4.5). As you take off the quantities from the plans, you list them on the spreadsheet. If you discover work items or materials during your preliminary study or during the takeoff that are not on your checklist, you can easily add them.

When taking off quantities, you can use two basic methods: the detailed method or the assembly method. In the detailed method, you simply list all of the individual items for each category from the checklist. The assembly method is another name for the component method discussed in Chapter 2. When using this method, you program your spreadsheet with the information to completely estimate the cost of a particular building component such as interior walls. The computer will help you generate all the items and costs associated with that component simply by entering that component's dimensions—such as length and height. Some software programs come with prebuilt assemblies as a part of the database. If you choose this option, you may need to modify the assemblies to conform to your method of building.

Advanced Options and Future Trends

After you become proficient with basic computer estimating systems, you may want to consider more advanced estimating tools that can further your capabilities.

Digitized Software

Using a digitizer is a fast, accurate way to measure and count. Many estimating software systems offer digitized software that you can use with digitizer hardware. A digitizer is simply an electronic board that is used with a stylus pen or pointer. The position of the pen is transmitted to the computer. It enables you to take off quantities straight from the drawings. You can take off lengths and areas and count parts. This information is not only stored in the computer but also may be displayed on the screen, thus allowing you to identify errors and omissions.

Sequence

Some software allows you to choose the sequence that you use to take off the quantities. You can view your work by different categories such as by material, by trade contractor, or by floor. This flexibility can be useful in entering your quotes and in scheduling your activities.

On-Line Pricing

Some material suppliers provide price lists electronically. In the near future you probably will be able to update your database for prices from almost any supplier within seconds.

Integrations with Computer-Aided Design

A close relationship exists between estimating functions, accounting, scheduling, pricing, and computer-aided design (CAD). Advances linking the estimating package with the CAD information are being made daily. Linking them electronically can save repetitive data entry and thus considerable time. Links between CAD and estimating can enhance the what-if analyses discussed earlier.

Packaged Programs

Several companies offer complete estimating packages that include productivity rates and pricing. Blindly using these systems can lead to serious problems. As mentioned, prices and labor rates vary by locale and time. If you are not willing to keep up with prices, you should hire someone to do so. Where as some packages have reasonable productivity rates, others do not. One such computerized package estimated over one-person year to erect and disassemble scaffolding around a one-story, moderately sized house on level ground. That is equivalent to using a three-man crew for four months to do nothing except erect and disassemble scaffolding on one house.

Summary

Using computer estimating systems can improve your speed, accuracy, and therefore, the quality of the decisions you make during the estimating process. If you invest not only the money but also the time to investigate, learn about, and implement computerized estimating systems, your work will show improvement.

Common Areas and Volumes

Area

Rectangle	$A = b \times h$	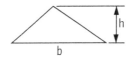
Triangle	$A = \frac{1}{2}b \times h$	
Parallelogram	$A = b \times h$	
Trapezoid	$A = \frac{(a + b)}{2} h$	
Circle	$A = 3.14\ r^2$	

Volume

Cube	$V = a \times a \times a = a^3$	
Rectangular solid	$V = w \times l \times h$	
Cone	$V = \frac{1}{3}\ \text{area of base} \times h$	
Pyramid	$V = \frac{1}{3}\ \text{area of base} \times h$	

Conversion Tables

TABLE 1 Concrete Quantities for Footings and Grade Beams

This table provides cubic yard per linear foot conversion factors for cast-in-place concrete footings and grade beams. The dimensions of the member can be indexed to quantities/linear foot (in cubic yards) and multiplied by the length of the member.

Cubic Yard per Linear Foot

Member width (inches)	Member depth (inches)						
	6	8	10	12	14	16	18
6	.0092						
8	.0123	.0165					
10	.0154	.0206	.0257				
12	.0185	.0247	.0309	.0370			
14	.0216	.0288	.0360	.0432	.0504		
16		.0392	.0412	.0494	.0576	.0658	
18			.0463	.0556	.0648	.0741	.0833
20			.0514	.0618	.0720	.0823	.0926
22				.0679	.0792	.0905	.1019
24				.0741	.0864	.0988	.1111
26					.0936	.1070	.1204
28						.1152	.1296
30						.1234	.1389
36						.1481	.1667

Waste = 5 percent (+ 5 percent if earthen formed footing)

Other factors can be calculated by multiplying the width (inches) by the depth (inches) by .0002572.

TABLE 2   Concrete Quantities for Slabs

This table provides square foot factors according to slab thickness that can be used to calculate the volume (in cubic yards) of concrete required. Divide the area required by the area covered to obtain volume.

	One Cubic Yard of Concrete	
Slab Thickness (inches)	Area Covered (square feet)
1	324
1¼	259
1½	216
1¾	185
2	162
2¼	144
2½	130
2¾	118
3	108
3¼	100
3½	93
3¾	86
4	81
4¼	76
4½	72
4¾	68

TABLE 3   Header Size

This table helps determine header sizes not shown on details. Select the building width and opening size and determine the member size. The length of the members is the opening width plus 8.

Header Size	Single-Story Building Width*			Two-Story Building Width*		
	22-24	26-28	30-32	22-24	26-28	30-32
2 ea. — 2 × 4	3'10"	3'7"	3'2"	2' 3"	2'	—
2 ea. — 2 × 6	6' 1"	5'7"	5'0"	3' 6"	3'1"	2'9"
2 ea. — 2 × 8	8'	7'5"	6'7"	4' 8"	4'1"	3'7"
2 ea. — 2 × 10	10' 3"	9'5"	8'4"	5'11"	5'2"	4'7"
2 ea. — 2 × 12	12' 5"	11'6"	10'2"	7' 2"	6'4"	5'7"

The header span is the maximum opening size that can be spanned by a double member. This is for lumber with allowable bending stress of 1,500 psi. For lumber with allowable bending stress of 1,000 psi, reduce allowable span by approximately 20 percent.

*The width is the dimension from one exterior bearing wall to the other exterior bearing wall.

TABLE 4 Reinforcing Bars

This table provides the weight/linear foot for each size bar to convert to pricing units. The lap factors are for continuous bars.

Designation	Diameter (inches)	Weight #/foot	Lap (40-bar diam.) 20" bar	30" bar	Factor 40" bar
2	0.25	0.167	.042	.027	.021
3	0.375	0.376	.063	.042	.031
4	0.50	0.668	.083	.055	.042
5	0.625	1.043	.104	.069	.052
6	0.750	1.502	.125	.083	.063
7	0.875	2.044	.145	.097	.073
8	1.00	2.670	.167	.111	.083

TABLE 5 Spacing Conversion Table

This table provides factors to determine the number of spaces required based on the designated spacing. The calculation will give the number of spaces required, and one (1) to obtain the number of framing members required. There is no allowance for waste, doubling of members or intersecting walls (for take-off of studs).

On-Center Spacing (inches)	Multiply Length by	Divide Length by
12	1	1
16	0.75	1.33
20	0.60	1.67
24	0.50	2.00
30	0.40	2.5
36	0.33	3

TABLE 6 Quantity Multipliers for Flooring, Siding, Wall Board

Multiply the net square footage of the area to be covered by the multiplier in the table to determine the *area* of material required.

| Type | Size | Multipliers | |
		Laid Straight	Diagonal
S4S	1 × 4	1.15	1.18
	1 × 6	1.14	1.17
	1 × 8	1.12	1.15
Ship Lap	1 × 6	1.27	1.30
	1 × 8	1.21	1.24
T & G	1 × 4	1.30	1.33
	1 × 6	1.21	1.24
	1 × 8	1.17	1.20
		Exposure	Multiplier
Bevel—1" lap	1 × 6	4½	1.35
	1 × 8	6½	1.25
	1 × 10	8½	1.20
Bevel—Rabbited	1 × 6	5¼	1.15
(Drop siding also)	1 × 8	7¼	1.10
	1 × 10	9¼	1.08

Conversion Area to Board Feet
1 × 4 area × 3   = bd. ft.
1 × 6 area × 2   = bd. ft.
1 × 8 area × 1.5 = bd. ft.
1 × 10 area × 1.25 = bd. ft.

TABLE 7 Roof Rafter and Sheathing Multipliers

To calculate the length of a rafter, take ½ of the building width plus the horizontal dimension of the overhang and multiply by the rafter factor listed in the table. The square footage of the sheathing can be determined by using the area factor. The valley or hip members can be determined by multiplying the horizontal dimension by the hip/valley factor.

Roof Slope	Rafter Factor	Hip/Valley Rafter Factor	Area/ Sq. Ft.
2/12	1.0138	1.4240	1.014
3/12	1.0308	1.4359	1.031
4/12	1.0541	1.4530	1.054
5/12	1.0833	1.4741	1.083
6/12	1.1180	1.5000	1.118
7/12	1.1577	1.5296	1.158
8/12	1.2018	1.5634	1.202
9/12	1.2500	1.6006	1.250
10/12	1.3017	1.6415	1.302
11/12	1.3566	1.6851	1.356
12/12	1.4142	1.7320	1.414
14/12	1.537		1.537
16/12	1.667		1.667

Lumber is purchased in 2' lengths, therefore calculations must be rounded up to even footage. Plywood sheathing square footage must be divided by 32 to determine the number of sheets required.

TABLE 8 Spacing to Units

Brick ties and other items are specified according to horizontal and vertical spacing requirements. This table provides conversion of spacing requirements to the number of pieces when the square foot of area is known.

Vertical Space	Horizontal Space	#/Sq. Ft.	Sq. Ft./Tie
12	8	1.5	.667
12	12	1.0	1.00
16	16	.057	1.75
24	16	0.38	2.67
30	16	0.30	3.33
36	16	0.25	4.00
24	24	0.25	4.00
30	24	0.20	5.00
36	24	0.167	6.00

Purchase unit is 100 or 1,000 lots for most items.

If noted as number of ties per square foot, or number of square feet per tie, the calculation is:

The total square feet × number of ties per square feet or
the total square feet + number of square feet per tie

TABLE 9 Roof Covering Factors

Add the proper percentage of roof area to allow for waste.

Roof	Wood Shingle	Asphalt Shingle
Shed/gable	8%	3%
Hip	12%	8%
Intersecting	12%	8%

Wood Shingle Coverage

Exposure to weather	Bundles/ square foot
4"	3.6
4½"	3.2
5"	2.88
5½"	2.62
6"	2.40
6½"	2.22
7"	2.06

Roof Sheathing Allowances for Waste
Change area to board measure.

Material	% Allowance
1 × 6 square edge	10
1 × 6 tongued and grooved	20
1 × 6 shiplap	20
1 × 8 square edge	12

TABLE 10 Concrete Quantities—Walls

This table provides cubic yard (and cubic foot) conversion factors for cast-in-place concrete walls. The square footage of the wall surface multiplied by the factor listed below will provide the volume of concrete required.

Wall thickness (inches)	Cubic Yards/ Sq. Ft.	Cubic Feet/ Sq. Ft.
3	.0092	.25
4	.0122	.33
6	.0185	.50
8	.0250	.67
10	.0310	.83
12	.0371	1.00
14	.0430	1.17
16	.0490	1.33
18	.0550	1.50
24	.0740	2.00
30	.0920	2.50
36	.1111	3.00
42	.1300	3.50
48	.1480	4.00

Waste = 5%

TABLE 11 Concrete Masonry Units—Quantities

This table provides factors to convert net square feet to concrete blocks and mortar. The factors are based on block laid in running bond with ⅜ inch mortar joints. The factor multiplied by the net square footage of the wall will provide the quantities required.

Blocks	Blocks/Sq. Ft.	Mortar/Sq. Ft. (cubic feet)
4 × 8 × 16	1.125	0.70
6 × 8 × 16	1.125	0.75
12 × 8 × 16	1.125	0.80
8 × 8 × 8	2.25	0.76
8 × 4 × 16	4.52	1.161
8 × 8 × 16	1.125	0.77

TABLE 12 Truss-Type Reinforcement—Quantities

Horizontal joint reinforcement is ordinarily specified as equally spaced. Determine the number of courses in the wall and subtract 1; divide the result by the factor listed in the table and round off to the next highest round number. Multiply the number by the length of the wall to determine the total linear footage of joint reinforcement required. The standard length of each truss is 10', and a 6"-lap is required. Add 5% for waste if openings are not deducted.

Horizontal Reinforcing Spacing	8"-High Course
8" o.c.	1
16" o.c.	2
29" o.c.	3
32" o.c.	4
40" o.c.	5
48" o.c.	6

TABLE 13   Clay Brick Quantities

This table provides factors to convert square footage of wall to brick and mortar quantities. Many estimators use a 5% factor for loss due to shipping breaks and on-site cutting.

Face Size Clay Brick	Running Bond Net Brick/sq. ft.	Brick with 5% Waste	Cuft Mortar with 25% Waste
Modular brick 2⅔″ × 4″ × 8″	6.75	7.00	.0588
SCR brick 2⅔″ × 6″ × 12″	4.50	4.75	.0988
Roman brick 2″ × 4″ × 12″	6.00	6.30	.0813
Norman brick 2⅔″ × 4″ × 12″	6.77	7.10	.0638

TABLE 14   Concrete Flock Fill—Quantities

This table provides the conversion factors to determine the amount of grout fill or loose insulation for concrete block cell and bond beam fill. The height of the wall multiplied by the factor for cell fill provides the quantity (in cubic feet and cubic yards) of fill required. The total length of the bond beam and lintels multiplied by the factor in the table for U-block fill factor provides the quantity of fill required.

	Horizontal Fill		Vertical Fill	
	cu. ft./ lin. ft.	cu. yd./ lin. ft.	cu. ft./ lin. ft.	cu. ft./ lin. ft.
Block (standard)				
6 × 8 × 16	.137	.0051	.137	.0051
8 × 8 × 16	.204	.0076	.024	.0076
12 × 8 × 16	.371	.0137	.371	.0137
U-Block				
6 × 8 × 16	.148	.0055	—	—
8 × 8 × 16	.216	.0080	—	—
12 × 8 × 16	.394	.0145	—	—

TABLE 15 Paint Covering Capability

Material	Square Feet Per One Gallon		
	1 Coat	2 Coats	3 Coats
Aluminum Paint	600	425	
Brick Paint—White or Light Tints on Unsurfaced Walls	225	110	75
Brick Paint—Dark Tints on Unsurfaced Walls	290	145	95
Enamel Base Coat (Prepared)	425	240	165
Enamels	425	215	165
Flat Wall Paint, Dark Colors (on Smooth Finish)	725	365	240
Flat Wall Paint, White or Light Colors (on Rough Sand Finish)	475	265	190
Flat Wall Paint—Sand Float Finish	300	150	
Flat Wall Hard Finish	500	300	
Flat—White Undercoat	500	300	
Linseed Oil	600		
Lacquer	200-300		
Lacquer Sealer	250-300		
Liquid Filler	250-400		
Non-Grain Raising Stain	275-325		
Oil Stain	300-350		
Outside House Paint—White or Light Tints, Porous Woods	475	255	
Outside House Paint—White or Light Tints, Close Grained Woods	575	300	190
Outside House Paint—Dark Colors, Greys, Tans, etc. Porous Woods	575	300	215
Paint and Varnish Remover (1 gallon should remove about 200 sq. ft.)			
Pigment Oil Stain	350-400		
Rubbing Varnish	450-500		
Stain, Wood Tints	750		
Stain, Shingle; (2 gals. to 1000 Shingles for dipping 1 coat. Brushing 1 coat after dipping, ½ gal.)			
Spirit Stain	250-300		
Varnish Stain	550	350	
Waterproof Paint	450	250	
Water Stain	350-400		

Caulking Compound: Coverage ½″ × ½″ Ribbon 77/in. ft./gallon.

TABLE 15 *Continued*

Paint Requirements for Interiors

Room Perimeter (feet)	Walls 8' ceiling (gals.)	8'6" ceiling (gals.)	9' ceiling (gals.)	9'6" ceiling (gals.)	Paint for ceiling	Finish for floors
30	⅝	⅝	¾	¾	1 pt.	1 pt.
35	¾	¾	¾	⅞	1 qt.	1 pt.
40	⅞	⅞	⅞	1	1 qt.	1 pt.
45	⅞	1	1	1⅛	3 pt.	1 qt.
50	1	1⅛	1⅛	1¼	2 qt.	3 pt.
55	1⅛	1⅛	1¼	1¼	2 qt.	3 pt.
60	1¼	1¼	1⅜	1⅜	2 qt.	3 pt.
70	1⅜	1½	1½	1⅝	3 qt.	2 qt.
80	1½	1⅝	1¾	1⅞	1 gal.	5 pt.

Each window and frame requires ¼ pt.
Each door and frame requires ½ pt.

TABLE 16 Nail Quantities

This table provides quantities of nails required for fastening lumber on carpentry items. Sum the quantities of lumber by type (in 1000 board foot measure—M fbm) and index to the table to determine the quantity, by weight, of nails required.

Item	Unit	Size	Pounds/ unit
Framing			
sills, plates, studs, joist, rafters	M fbm	10d common	6
Exterior Sheathing			
1 × 4	M fbm	8d	48
1 × 6 matched	M fbm	8d	32
1 × 8	M fbm	8d	27
1 × 10	M fbm	8d	20
gyp bd.	100 sq. ft.	4d-½"	1.5
plywood	100 sq. ft.	6d (or ed)	1
Flooring			
1 × 3 softwood	M fbm	2½ brads	32
1 × 4 softwood			26
1 × 6 softwood			18
hardwood	M fbm	1½ flooring	12
		2¼ flooring	24
Interior Sheathing			
gyp wallboard	100 sq. ft.	5d ring shank	1.0
plywood paneling	100 sq. ft.	3d finish	12
matched wood	100 sq. ft.	6d finish	6
Roofing			
asphalt shingle	square	1" galvanized	3.0
wood shingle	square	4d shingle	3.0
Casing and Base	100 lin ft.	6d casing	1.0
Finishing Boards	M fbm	8d finishing	0.25
Furring and Masonry	100 lin ft.	1" masonry	1.0

C

Enlarged Floor Plan

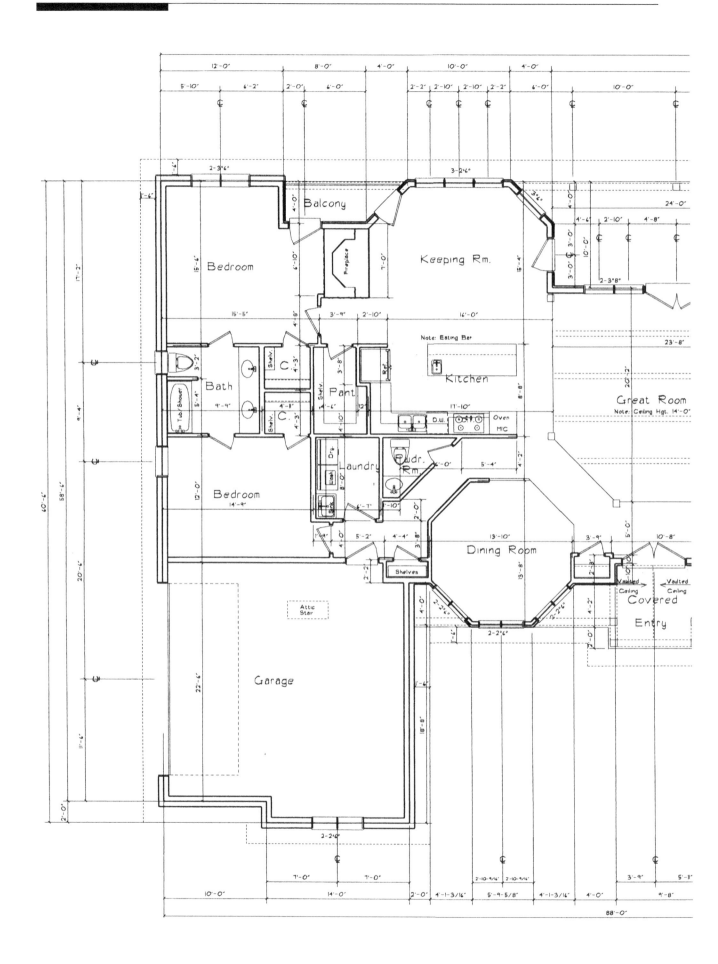

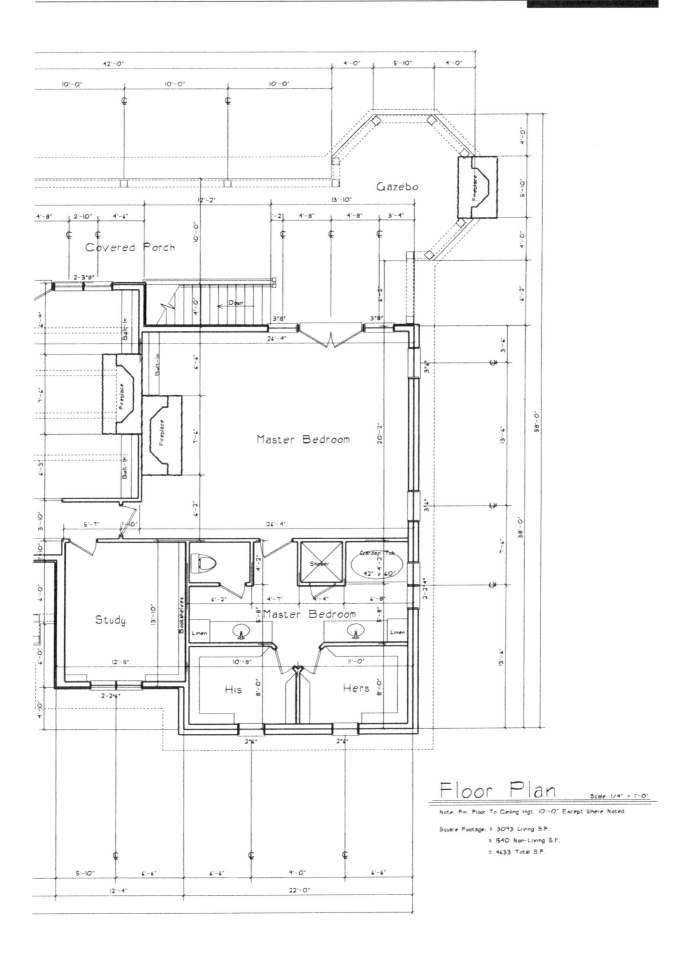

Gazebo

Covered Porch

Master Bedroom

Study

Master Bedroom

His   Hers

Floor Plan   Scale: 1/4" = 1'-0"

Note: Fin. Floor To Ceiling Hgt. 10'-0" Except Where Noted.

Square Footage: ± 3093 Living S.F.
                ± 1540 Non-Living S.F.
                ± 4633 Total S.F.

Enlarged Front
and Rear
Elevations

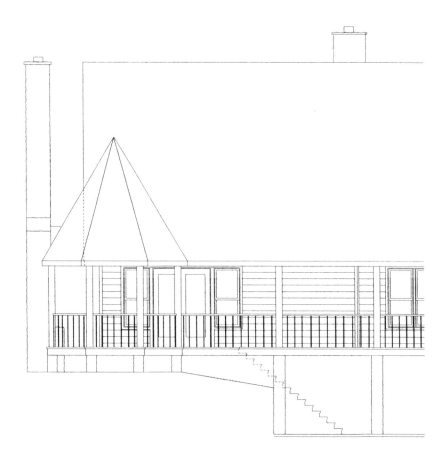

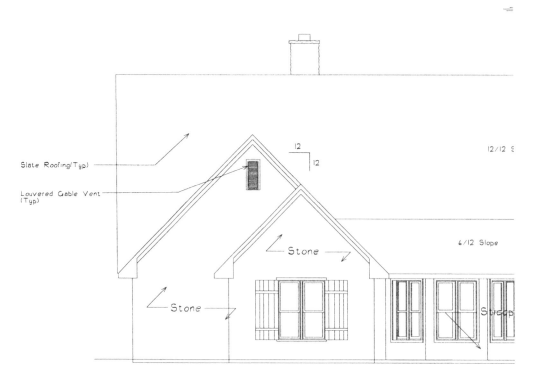

Slate Roofing(Typ)

Louvered Gable Vent
(Typ)

12
12

12/12 S

4/12 Slope

Stone

Stone

Front

Rear Elevation   Scale: 1/4" = 1'-0"

Stucco   Stone

12
12

Slope

Elevation   Scale: 1/4" = 1'-0"

Glossary

boards—Lumber cut less than 2 inches in nominal thickness and 2 or more inches in nominal width.

black-in—The point at which the house has felt on the roof and sheathing on the walls.

brick veneer—The nonstructural, outside facing of brickwork used to cover a wall built of other material.

checklist—A listing of takeoff, subcontract, and cost estimate items used to organize and arrange the takeoff and cost estimate.

check-off—The method of using a checklist that requires each item shown on the working drawing to be checked off as it is counted or measured for the quantity determination and taken off for the cost estimate.

contingency cost—A cost that is added on when total project conditions cannot be anticipated nor estimated. Without a contingency cost, these unanticipated costs are offset by either the margin or unanticipated cost savings in other areas.

contract documents—The agreement between the builder and the home buyer including conditions, working drawings and specifications, addenda, modifications, and any other papers of official agreement.

cost estimates—Quantity takeoff summaries for work items listed in appropriate pricing units for figuring cost extensions to complete the estimate of labor, material, and/or equipment costs required.

cost extensions—The mathematical computations required to convert the quantity takeoff into a cost estimate. It requires multiplying the unit costs by the quantities determined and adding the individual cost estimates for each category (such as labor, material, and equipment) on each cost estimate form.

cost overrun—The amount by which the actual cost of an item exceeds the cost estimated for that item.

cubic foot—A 1 × 1 × 1-foot unit of volume. (A cubic yard contains 27 cubic feet.)

detail—An enlarged drawing of a part of another drawing that indicates precisely the design, location, composition, and relation of the elements and materials shown.

dimension lumber—Lumber that is at least 2 inches up to but not including 5 inches in nominal thickness and 2 or more inches in nominal width. Dimension lumber may be classified as framing, joist, planks, rafters, studs, and small timbers, and typically they are 2×4s, 2×6s, 2×8s, 2×10s, and 2×12s.

elevation—A two-dimensional, geometrical graphic representation of a building or object on a vertical plane—a picture view included in the contract documents.

equipment and jobsite overruns—The cost incurred when a project's duration and/or the use of a piece of equipment exceeds the estimate.

equipment and jobsite overhead contingency cost—A cost added as a percentage of the total equipment and jobsite overhead costs for the project. Some estimators increase the estimated time the builder will use the equipment or figure the rental rates higher than they will be.

equipment quantity—Equipment use measured in short-term rental rates (daily or hourly) for equipment used for a limited time. Long-term rental rates (weekly or monthly) are used for equipment required for longer duration (possibly on multiple sites that share cost).

estimate summary—A list of the total labor, material, equipment, subcontract, and overhead costs. It often includes the computation of labor burden, sales tax, and markup or profit.

estimate worksheet(s)—A listing of measurements, computations, and material quantities determined from a set of plans. The worksheet must have a title block that includes the name of the project, the number of the page, and the type of work covered on that page. Each worksheet and the sequence of worksheets should be organized according to the checklist for each takeoff and estimate.

furr down—Build down.

furr out—Build out.

jobsite overhead—Those costs for supervision, temporary utilities and facilities, layout, building permits, bonds, and insurance that are listed item-by-item and segregated by method of payment, including those computed on duration of use and those figured as a percentage of direct cost estimates. The cost extensions include labor, miscellaneous costs (for example, layout), materials, and equipment.

indirect costs—The cost of labor burden and sales tax. Units for measuring labor burden are usually a composite percentage of direct labor costs. The sales tax is computed as a percentage of total material costs.

labor contingency—This cost is figured in anticipation of varying productivity in the building of the house. Builders often adjust the standard unit costs that they apply to the listed quantities of work to account for jobsite conditions. Unpredictable conditions, such as weather, affect labor productivity.

labor cost overrun—The amount by which the labor cost exceeds the labor cost estimate. Such overruns result from inaccurately calculating the quantity of work, an optimistic estimate of crew productivity (a higher actual unit cost), or increased labor rates.

labor quantity—An amount of labor measured on an hourly and a unit-cost basis. The units of purchase for unit-cost labor will equal the labor-only units used for items in trade contracts, including square footage of the building, roof, finish, or number of pieces (masonry). For labor bought on an hourly basis, labor costs depend on labor productivity rates from historical cost records. These records are expressed in the units of measurement used in cost control.

linear foot—A line 1 foot in length.

lump sum—A total cost for all of the work or material required to complete a segment of work for a house.

markup—The amount of money included in the estimate in addition to the direct costs of trade contracts, material, labor, equipment, and jobsite overhead. Some builders use a percentage of the subtotal cost for calculating markup. Other builders use a method of annual general overhead compared to the annual volume of work expressed in dollars or number of projects.

material contingency cost—An amount of money included in the estimate in anticipation of some loss of materials. Standard practice includes a percentage for loss in converting measurements or quantities to pricing units or as an addition to the number of units to be purchased.

material cost overrun—The amount by which material cost exceeds the estimate of that cost. These overruns occur when a price escalates or when the quantity of materials purchased exceeds the amount the estimate allows.

material quantity—An amount of material measured in the same units as the units of purchase from vendors, including the amount of an item, board feet, and/or number of pieces by size, length, square feet, linear feet, sheet, or square.

plan view—A depiction of a horizontal section of a building or object through the walls. This contract document shows such items as openings, recesses, projections, and columns.

price—The current valid cost of a particular material if the material is bought at the time of inquiry.

price escalation—An increase in the price of a material between the time of inquiry (estimate) and the time of purchase.

price-escalation contingency cost—An amount of money added to the estimate to cover potential price increases for materials, usually as a percentage of the total of prices used in the estimate.

pricing units—The units of measurement used to express the quantities of materials on the quantity takeoff. Unit costs must be in the same pricing units as the quantity takeoff.

quantity—The amount of material, labor, equipment, or item of work required for one cost category of a house.

quantity determination—The measurement of individual items from the working drawings and computations required to determine a quantity of material or work required.

quantity takeoff—An organized arrangement and listing of quantity determinations arrived at by reviewing plans and specifications and measuring quantities.

quotation—A guaranteed price or bid, often with a time limit, offered by a vendor or trade contractor for a specific material or segment of work. Quotes are preferred because they are not subject to escalation and do not require contingencies.

schedule—A timetable for completing a job; also a list or table of parts including doors, windows, and room finishes.

section view—A drawing of an object as if it were cut lengthwise to show the interior makeup; often a contract document.

square foot—A 1- × 1-foot unit of area.

studs—A series of slender wood or steel members (typically 2×4s or 2×6s) used for the structural and nonstructural walls and partitions.

timbers—Lumber that is nominally 5 or more inches in at least one dimension. Timber may be classified as beams, stringers, and posts, usually they are 6×6s and 8×8s.

trade contract bid form—An organized format for recording and comparing trade contract bids and quotes. A similar form can be used to compare material quotes.

trade contract contingency cost—A cost added to lump-sum trade contract bids only when a bid is perceived or validated to be too low. Trade contracts based on unit cost require builders to determine the actual billed units in their computations.

trade contract cost overruns—The amount of extra cost incurred for unit-cost contracts when estimated units do not equal actual measured units. When a trade contractor defaults on a bid, the unit or lump-sum cost of the work often exceeds the amount included in the cost estimate.

trade contract estimate form—A summary page of trade contract items formatted to list the trade contractors' bids and quotes used in the cost estimate. A similar form can be used to list and summarize the bids and quotes of material vendors.

unit costs—The estimated or quoted cost for a particular item of material or work per standard unit of measurement. The pricing unit used for cost must be the same as that used for the quantity takeoff.

Resources & Recommended Reading

Christofferson, Jay P. *Estimating with Microsoft™ Excel*, 2nd ed. Washington, DC: BuilderBooks, National Association of Home Builders, 2003.

_____. *EstimatorPRO™*, 5.1. Washington, DC: BuilderBooks, National Association of Home Builders, 2005.

Cough, Richard H., et al. *Construction Contracting: A Practical Guide to Company Management*, Hoboken, NJ: John Wiley & Sons, 2005.

Crump, David, and David Jaffe. Contracts and Liability, 5th ed. Washington, DC: BuilderBooks, National Association of Home Builders, 2004.

Johnson, Kenneth V. *Building Spec Homes Profitably*. Kingston, MA: Robert S. Means, 1995.

Means Residential Cost Data 2005, 25th ed. Kingston, MA: Robert S. Means, 2005.

Ogershok, David, and Richard Pray. *2006 National Construction Estimator*, 54th ed. Carlsbad, CA: Craftsman Book Co., 2005.

Paxton, Albert S. 2006 *National Repair and Remodeling Estimator*, 28th ed. Carlsbad, CA: Craftsman Book Co., 2005.

Peurifoy, Robert L, and Gerold D. Oberlander. *Estimating Construction Costs*, 5th ed. New York: McGraw-Hill, 2001.

Siddons, Sott, ed. R. *Walker's Building Estimator's Reference Book*, 27th ed. Lisle, IL: Frank R. Walker Co., 2005.

Shinn, Emma. *Accounting and Financial Management*, 4th ed. Washington, DC: BuilderBooks, National Association of Home Builders, 2002.